THE GOVERNING OF AGRICULTURE

STUDIES IN
GOVERNMENT AND PUBLIC POLICY
Charles H. Levine, Series Editor

BRUCE L. GARDNER

THE GOVERNING OF AGRICULTURE

PUBLISHED FOR THE MANHATTAN INSTITUTE
FOR POLICY RESEARCH AND THE INSTITUTE
FOR THE STUDY OF MARKET AGRICULTURE
BY THE UNIVERSITY PRESS OF KANSAS

Second printing, April 1983
Published by the University Press of Kansas (Lawrence, Kansas 66045),
which was organized by the Kansas Board of Regents
and is operated and funded by Emporia State University, Fort Hays State
University, Kansas State University, Pittsburg State
University, the University of Kansas, and Wichita State University
ISBN 0-7006-0215-1 Paper
Library of Congress Catalog Card Number 81-50692
Printed in the United States of America

To Mary

Contents

Foreword

It is rare to come upon a book dealing with the public interest in food and agriculture that is as clear and competent as this book is in clarifying the consequences of public choices that we as citizens make. Information that can be trusted is hard to come by. Citizens, including students and scholars, are bombarded by distortions of information. They are assailed by a flood of private pamphlets, public documents, and books that distort the public interest. This book is indeed one of the exceptions. It is marked by high quality of workmanship in dealing with the complexities of many different public programs in this area. Although there are no simple solutions for the unsettled issues, the presentation is remarkably free of professional jargon.

The author does not make the mistake of lumping the many complex public programs dealing with food and agriculture; he considers both the political process and the economics of several important classes of such programs. For each class the available evidence is interpreted with care. The author is meticulous in not claiming too much for the evidence on which he is dependent. Anyone who invests a few hours of his time in reading this book will see clearly the various effects of his choices as a citizen, and he will be well repaid.

What distinguishes this book can be stated briefly.

It is committed basically to education, not to quick solutions which are impossible in view of the conflicting special interests and the prevailing controversy on the issues. Education that increases the understanding of the electorate and also of students and scholars of these public choices becomes, at best, worthwhile only over the long pull.

It is also an intellectual commitment to the proposition that both the political process and economics matter and that each must be treated seriously.

The book is distinguished by its qualitative thinking and its quantitative analysis.

What I find especially rewarding is the classification of public programs: for example, the clarity with which the effects of the tobacco and sugar programs are revealed. They have much in common. The programs that are designed to alter the supply tell quite a different story from those that seek to change the demand. Then there are the effects of the marketing orders on consumption and production. But not least in importance are the estimates of what consequences these various public food and agricultural programs have for the consumer, the producer, and the taxpayer.

Theodore W. Schultz

University of Chicago

Preface

This book is about the public interest in providing the nation with food and fiber. It is concerned particularly with the consequences of the United States government's policies regarding agriculture and food. Is the public interest being well served by these policies? and if not, What alternatives would do better?

My answers are that our policies do not promote the public interest as well as they should and that the principal route to improvement is through less intervention in the commodity markets. This is not, of course, a novel point of view. There is a need, however, to show that reduced intervention is appropriate for the U.S. agricultural situation of the 1970s and 1980s. Obviously, this can't be demonstrated as an indisputable inference from facts and logic. But I believe that there are good reasons for adopting the views put forth, and my goal is to state these reasons as clearly as possible.

The book is organized as follows. Chapter 1 provides necessary background information about U.S. agriculture. Chapters 2 and 3 present the main elements of the government's current agricultural policy. In chapter 4 the meaning of "the public interest" is discussed, and current policies are evaluated. Chapter 5 considers criticisms of this evaluation, highlighting "market failures" in U.S. agriculture. Chapter 6, which discusses the potential for public policy to remedy the deficiencies of past programs and of the market, highlights "government failures."

The second half of the book, chapters 4 through 6, is explicitly evaluative, although I have not hesitated to include value judgments in the first half also. In this sense the book is more nearly a tract than a research study. My approach is essentially critical. It begins with things as they are, then looks for flaws, and finally tries to suggest improvements. It may be thought that while it is an interesting exercise to criticize the existing state of affairs, current farm policy exists for good and sufficient political reasons, so that criticism is, in a practical sense, pointless. I believe that this view is mistaken. In many respects, our current farm programs are better than their predecessors because we have learned from experience—an essentially critical activity. Therefore, I believe that criticism of current policies, if it tries to be fair and if it is factually and analytically well based, has a useful role to play.

There are many more ways to be wrong than to be right in evaluation of policies. Nobody could hope to put forth only correct judgments, convincingly stated, where there is so much analytical uncertainty and divergence of interests. From the most general level—what counts as "in the public interest"—to the analytical details of how particular governmental actions affect the food sector, there are conflicting points of view. And often a plausible case can be made for more than one of them.

Elimination of some of the least convincing parts of earlier drafts of this material has been aided by the criticism of several individuals. I would like especially to mention the comments of my Texas A&M colleagues Michael Cook, James Richardson, and Ron Knutson. They do not, of course, bear any responsibility for unconvincing and erroneous content that may remain or for the views expressed.

1

The Present State of Farming and Farmers in the United States

As background for informed judgments about the governing of agriculture, it is necessary to be acquainted with some essential facts about the nature of farming, the economic status of farmers, and the history of the government's intervention in U.S. agriculture. The first order of business is to isolate the features of food consumption and production that make the agricultural sector a plausible candidate for special treatment by the government.

THE ECONOMICS OF MODERN AGRICULTURE

What is special about agriculture? One of the chief impediments to clear thinking about farm policy is the superfluity of answers to this question. In debate on policy issues, one often hears: "Sugar is unique"; "The cattle business is a special case"; "Milk is different." Such assertions, as well as detailed arguments to back them up, are put forth in abundance by lobbyists in Washington, D.C., when they are supporting the cause of the commodity producers whom they represent. Not only is agriculture special, but so is each agricultural product. The solutions proposed, however, are basically the same in each case: governmental aid to producers in the face of a threat to their economic well-being.

The facts that arguments to this effect are used in special pleading and that some of them are specious should not make us so skeptical that we refuse to pay attention to the particular characteristics of agriculture.

It has been said that agriculture is the oldest economic activity of civilized man—the activity which, in fact, defines the dawn of civilization. Modern agriculture retains one important feature of its ancient heritage: its closeness to nature. This is reflected in the sensitivity of farm output to weather, insects, disease, and other natural hazards. It is reflected also in the biological nature of agriculture production processes, which results in lags and periodicities that are not found in nonagricultural industries. The result is that agricultural markets are characterized by random shocks which generate short- and long-term quasi-cyclical behavior of prices and output.

While each agricultural production process has its unique and complex characteristics, these features do not negate basic economic principles. Producers still respond to opportunities to earn greater returns, so that higher prices call forth larger quantities of goods, and lower prices, smaller. Similarly, it is still true, on the consumer side, that despite the stomach's insistence on having something to fill it and despite our peculiar desires for particular foods, when food prices rise, less is consumed, and when food prices fall, more is consumed. Indeed, even when people are living at a bare subsistence level, consumption responds to the price of food. The reason is simply that if a family is at the margin of subsistence, it is almost by definition consuming necessities only. In a poor country, food may account for as much as three-fourths of a poor family's income. Under these circumstances, if the price of food rises by 20 percent, the family simply must cut back on food.

Not only do food consumption and farm production respond as one would expect from elementary economic theory, but the relationships are in many cases so stable that one can estimate them with some reliability. Indeed many of the econometric techniques that are now so widely used in explaining events in sectors of the economy were first developed and applied with reference to agricultural products.

In the "explosion" of food prices that was associated with the grain sales to the Soviet Union and with short crops during the period 1972 to 1974, the real price of food (as measured by the consumer price index [CPI] for food, deflated by the overall CPI) rose 11 percent between 1972 and 1974, while real expenditures on food, deflated by the food CPI, declined by about 3 percent. This would imply an elasticity of demand for food of –3/11, except that during this same two-year period, the population increased by 1½ percent and real income per capita increased by 2½ percent. After adjusting for the increase in food consumption that these factors could be expected to cause, the data suggest an elasticity of demand for food in the neighborhood of –.4. Econometric estimates typically indicate that the demand response to prices is a little less than this for food as an aggregate commodity. For individual commodities, estimates of elasticities range from around –0.1, for staples like cereals and sugar, to the neighborhood of –1.0, for meat products, with demand being even more elastic for some specialty items.

On the supply side, how production will respond to price incentives is even more difficult to estimate with precision because of the randomness of output and because of time lags. Farmers must therefore base their production plans on expectations of prices at a future date, expectations that are not directly observable. Nonetheless, there is strong evidence that farmers in the U.S. do respond positively to price signals and that the responsiveness is perhaps greater in the United States than in most other countries. For example, comparing the 3-year periods before and after the 1972/73 shocks to the international grain markets, world grain acreage was about 4 percent higher in the latter, whereas grain acreage in the U.S. increased by 12 percent. The difference between the U.S. and the foreign production responses to price is probably in large part attributable to different policy regimes, because many countries attempt to control domestic prices rather than allowing them to fluctuate with world market prices. In any case, it seems clear that supply does in fact respond to price in the United States.

The significance of the fact that demand and supply respond to prices for agricultural commodities is elementary but fundamentally important. It is, first, that expressions such as "food shortages" or "surpluses" only make sense in reference to a particular market price. There is no real question of an absolute insufficiency of food, although the potential for a rising real price of food is a serious issue. A corollary is that there is normally only one price that does not result in either surpluses or shortages. Over and over again, governmental action with regard to agriculture has been undertaken out of dissatisfaction with prevailing market prices. And over and over again, these actions have come to grief because it has been found that the attempt to establish different prices has created either surpluses or shortages.

The relative inelasticity of both supply and demand, coupled with the substantial random element in production, creates the well-known chronic instability in prices for farm commodities and hence in the incomes of farmers. It is possible that public policy can promote everyone's interests jointly, rather than merely shifting dollars from one set of pockets to another, by providing remedies for instability. However, the issues here are complex, and it is by no means a foregone conclusion that government is capable of delivering these public benefits.

A final idiosyncratic feature of the agricultural sector is that it is in reality two industries, not just one. There is a crop-producing sector—the land-based agricultural industry, which combines the sun's energy with land and water in order to produce human nutrients—and a quite distinct livestock industry, which uses crops as feed inputs. It is true that many agricultural firms are vertically integrated in that they produce for both industries; but while this is a reason for considering the two together, there are real difficulties in doing so. Because the first industry's output is an input into the second industry, one must exercise care in aggregating the two. It makes no more sense to add up the value of corn and of livestock production to get the value of "agricultural output" than it would to add up the value of automobiles and of the steel used in them to get the value of "car-and-steel output." More important for purposes of policy analysis, livestock producers are consumers of grain, and in many policy issues, they have interests that are parallel to those of the final consumers of food. These conflicting interests are often blunted in their political impact because so many farmers are producers both of crops and of livestock. Nonetheless, the economics are essentially the same as they would be if the industries were separately owned and operated. For example, the sharp increases in prices for grain during the early 1970s, while they constituted a bonanza to the crop-production industry, played a substantial role in throwing the livestock-production industry into disarray from which it had not fully recovered five years later.

Modern agriculture has progressed economically because capital has been infused into the production processes. This capital includes machinery to enhance the productivity of labor; chemical and biological innovations to ensure more output from given inputs; transformations of land from less to more productive conditions—for example, by drainage or irrigation; and improvements in the skills of farmers themselves through investment in human capital. As a result of the modernization process, a diverse and complex agriculture has evolved within the United States, and even more dramatic differences have arisen between the developed countries and the poor countries in which the modernization of agriculture via nontraditional investment is just beginning.

The growth of agricultural productivity has resulted in a trend of declining real prices for agricultural commodities. Relative to the overall CPI, the prices of farm products in 1980 were about 40 percent lower than they were in 1950. The relative price declines have given rise to calls for "parity," the idea that prices for farm products should bear a fixed relationship to farm input prices in order to keep farmers' incomes from declining. However, when given inputs generate more output because of growth in productivity, declines in product prices can easily give false signals about farmers' incomes. While there is no doubt that consumers are the prime gainers from growth in productivity which reduces commodity prices, it does not follow that the producers are therefore losers. Growth in productivity is not a zero-sum game.

Because the well-being of farmers in modernized U.S. agriculture is a major issue in farm policy, it is important to know how producers are faring economically. This question cannot be answered by a consideration of production and prices alone, however. It requires other information, to which we now turn.

THE ECONOMIC STATUS OF FARM PEOPLE

Farmers are an extremely diverse group. The core idea of a farm is an enterprise that is family owned and operated. The chief executive officer is also the day-to-day manager and laborer, the bearer of risks, and the owner of at least some of the capital or land in the business. How far an enterprise can stray from this core idea and still properly be labeled as a farm (or a ranch) is a matter of controversy. The definition is even politically controversial, because the disbursement of some federal funds to states depends on the number of farms in each state.

As one becomes more inclusive in counting farm people, the ability to make generalizations about their economic status decreases. Some individuals own farms but live in towns or cities. Such individuals, who never or seldom even visit their farms but merely rent them to others, are landlords, but not farmers. What if they visit their farms every day, make some operating decisions, but have a share tenant on the farm? Then the official U.S. statistics call the tenant, not the landowner, the farmer. The general rule is one farmer per farm. Other individuals who live in rural areas have small agricultural enterprises but obtain the bulk of their incomes from nonfarm jobs. They are counted as farmers, and their enterprises are counted as farms if they sell $1,000 per year of agricultural products. Since $1,000 of product can easily be obtained from, for example, one dairy cow or five acres of corn or one-half an acre of tobacco, many part-time operations that have no commercial significance are counted as farms.

Table 1.1 provides some indicators of the economic status of farms, classified by value of sales. The mean income per farm in 1978, as estimated by the United States Department of Agriculture (USDA), is about $23,000, including both farm and off-farm sources. The Census Bureau's Current Population Survey (CPS) for 1978

TABLE 1.1
INCOME AND WEALTH OF FARM OPERATORS, 1978

	All Farms	Sales Class			
		Less than $2,500	$2,500 to $39,999	$40,000 to $99,999	$100,000 and over
Number of farms (in thousands)	2,672	916	1,179	390	187
Percentage of all farms	100	34	44	15	7
Income from farming ($ per farm)	10,037	1,738	5,936	21,636	52,337
Off-farm income ($ per farm)	12,829	17,205	11,721	6,846	10,850
Total income	22,866	18,943	16,657	28,482	63,187
Net worth ($/farm)*	259,000	107,000	185,000	384,000	1,038,000

SOURCE: United States, Department of Agriculture, *Farm Income Statistics* and *Balance Sheet of the Farming Sector.*

* Proprietors' assets minus debts as of January 1, 1979 (estimated).

estimated the median money income of farm families at $15,300, compared to $17,700 for nonfarm families.[1] Making allowance for the likely underestimation of self-employment income in the CPS and for the favored tax treatment of farm income, the average after-tax income of farm families is probably no less than the average income of nonfarm families.

Table 1.1 indicates that the definition of a farm has become too inclusive to be meaningful as an indicator of well-being for people who are engaged in agricultural production as a means of livelihood. The table classifies as a farmer any rural resident who sells more than $250 of products, a lower boundary point than the $1,000 mentioned above (the latter being a new definition that has not yet been incorporated in the economic statistics). Note that on the smallest "farms," which constitute 34 percent of the total number, more than 90 percent of family income is received from off-farm sources. These farms account for only about 1 percent of total U.S. sales of farm products.

If one is interested in the people who actually produce the nation's food and fiber, it is best to restrict attention to farms that have $40,000 or more in sales. These roughly half-million producers accounted for over 80 percent of U.S. farm sales as of 1978. The $40,000 level does not by any means indicate that our view is restricted to gigantic operations; $40,000 is approximately the value of output from 30 dairy cows, 150 acres of cotton, 200 acres of soybeans, or 400 marketed hogs.

Commercial farmers and their families are not an economically deprived group. Their combined farm and off-farm incomes average more than twice the amount of the average income per family in the United States. Moreover, a comparison of after-tax incomes would show that farmers have still greater advantages. Internal Revenue Service (IRS) statistics show farm profits subject to tax of $3.5 billion in 1975 and 1976 and $0.5 billion in 1977.[2] Average taxable farm profits were $2.5 billion for this three-year period (excluding profits of corporations in farming). The USDA's estimates of aggregate net farm income averaged $21 billion during the period 1975 to 1977. Assuming that 15 percent of this total was received by corporations or partnerships that did not file individual income tax returns or was received in the form of realized capital gains that were not reported as farm profits for tax purposes, the actual farm income corresponding to the $2.5 billion reported by the IRS was $17.9 billion. Thus, about 86 percent of farm income appears to have been sheltered from federal income taxation. The main reason seems to be more liberal deduction of expenses than is available for most nonfarm salaries or business income. Another reason is that the USDA statistics include income in kind, whereas IRS statistics do not. If the tax rate on the sheltered $15.1 billion had been an average of 25 percent, the tax break to U.S. farmers amounted to $3.8 billion per year, or $1,400 per farmer.

Table 1.1 also shows data on the value of farmers' assets minus debts—the net worth owned by farmers. Nonfarm people accrue capital gains also, notably through home ownership, but the magnitude per family is much greater for farmers. Commercial farmers, those in the $40,000+ sales class, are in fact considerable capitalists, possessing a mean net worth, in farm and liquid financial assets, of $500,000 per farm. This figure excludes nonfarm assets owned by farmers, the value of which cannot be precisely measured but which must be substantial.[3]

The preceding description of farm prosperity is not meant to obscure the fact that there is great diversity in the economic condition of farmers. Those who specialize in commodities that experience low prices, such as wheat in 1977 or cattle from 1975 to 1977, can suffer severe losses, even facing bankruptcy in some instances. These cases, however, are the exception, even more so than in nonfarm business enterprises. In 1977, 735 farm operators in the United States filed for bankruptcy, the greatest number in recent years.

This amounts to about 1.5 per 1,000 operators of commercial farms. During this same year, there were 6,533 bankruptcies among merchants, 2,680 among professionals, and 214,399 in the U.S. as a whole. Farmers had a lower rate than had any other business or professional group. It may be, however, that many farm businesses fail without going through formal bankruptcy. This also seems to be true for small nonfarm business. In terms of the broader indicators of financial difficulty that are available, such as delinquencies in loan repayments, farmers again appear to be less prone to failure than are nonfarm business firms. In 1977/78, there were 2,400 foreclosures on farm mortgages in the United States.[4] But there are over 2 million farms. "Many of our farmers are on the verge of bankruptcy,"[5] or words to that effect, have led U.S. policy makers down some dubious pathways, so it is worth keeping in mind how little truth there is in such words as a distinguishing feature of U.S. agriculture.

A large number of rural-farm families fall below the poverty line. The incidence of poverty, as defined by the Bureau of the Census, among rural-farm families averaged 14.1 percent in the period 1975 to 1977, compared to 9.5 percent for nonfarm families.[6] But as recently as 1959, 43 percent of rural-farm families fell below the poverty line. This rate of improvement—a reduction of two-thirds in the incidence of poverty in less than twenty years—far outstrips the progress made in fighting poverty in the economy as a whole. It is not developments in commercial agriculture or farm programs, however, that have pulled rural people out of poverty or have caused the poverty that remains. The rural poor are in the subset of those who produce small amounts of farm products, principally the category of less than $10,000 in sales, who do not have the off-farm sources of income that most of the families in this category have. Even very large increases in prices for farm commodities would not pull these families out of poverty.

Farming is a risky business. A farmer's production depends on the weather and on other unpredictable factors. Because the demand for farm products is inelastic, fluctuations in output generate severe instability in prices and income. But the really problematic feature of agricultural instability is not so much year-to-year variations as the longer-term periods of low returns. Annual variation is transformed to longer-term periodic fluctuations by commodity storage (for crops), by the length of the process of biological production (for livestock), and by lags in adjusting the allocation of fixed capital and skills. Adjustment problems are complicated by continuing structural shifts in the production and demand sides of markets for farm commodities, which make it impossible to say with assurance whether an observed year of changed conditions is permanent or transitory. The risks are magnified by the necessity for large amounts of capital in modern farming. Unless a farmer inherits wealth, he must early in his career assume quite massive debts. The debts that are incurred in order to purchase fixed capital equipment and land result in extremely variable net returns to the operator's equity.

In addition, the farmer today is much more subject than most of us are to risk due to variation in the rate of inflation and to changes in the regulatory environment. Nonfarm businesses face many though not all of these risks. But they are, in enterprises of any size, typically protected by the laws of limited liability for corporations. The farmer is literally gambling for his family's standard of living. It is as if a worker had to play roulette to determine the size of his paycheck, and as if zero paychecks for a period of several years were a real possibility.

To put farm earnings into perspective, a few more facts are also pertinent. First, in part because of the economic uncertainty that they face, farmers have much higher rates of saving than the general population. They thus tend, over their lifetimes, to accumulate wealth, and a substantial part of their income represents a return on this investment. Thus, the higher incomes of commercial farmers do not represent high rates of returns on resources so much as returns on savings and investment that they have made over the years. Second, farmers are legendary as being hard workers. They put in long hours under sometimes stressful and unpleasant conditions. People in business or other occupations who work as hard and save as intensively also earn rather more than average U.S. incomes.

In short, it would be unfair to conclude that operators of commercial farms are being overpaid. It would also be premature to conclude that just because commercial farmers are, as a group, not badly off economically, any intervention in commodity markets to assist farmers is unjustified. What the data on economic status do indicate is that there is no case for farm commodity programs *as an antipoverty policy*. The people who will be helped by such programs are not a class of poor people; and while there are still disproportionate numbers of poor people in agriculture, they will not be removed from poverty by commodity programs.

It is important to emphasize these points because of a tendency of political spokesmen for commercial-farm interests to paint a picture of oppressed farmers, who are subjected to unfair treatment from all sides and who are on the verge of bankruptcy. One reason for the widespread acceptance among nonfarm people of governmental aid to farmers is a too uncritical acceptance of this misleading view.

There is, however, a very significant new economic problem confronting some farmers today. It is reflected in the following remarks made recently by the secretary of agriculture: "While we are telling farmers that they are well off—that farm prices, net farm income, and farm exports are up dramatically—some farmers continue to tell us that in the midst of stability and prosperity, they can't make it. Individuals have individual problems—problems they explain in terms of machinery and land and investments and debt—and averages do not address their very real and specific situations."[7] The secretary identifies these problems with the "structure of agriculture," the losers being the smaller farms. While individuals or regions whose production costs are particularly high undoubtedly have special difficulties, the real problem that the secretary is hearing about is more likely a cash-flow problem that affects large as well as small operators.

This problem, in its new and virulent form, is a consequence of a high rate of inflation. Consider Mr. A, who wants to purchase 200 acres of farmland for $1,500 an acre in order to enlarge his farm. Having an equity of $100,000 in cash and land, he can borrow the $300,000 needed for the purchase, paying a 9 percent interest rate. So Mr. A, after the purchase, has assets of $400,000, with a $300,000 debt to go with his $100,000 total equity. Mr. A's income statement for this land will look something like this: the $400,000 in land yields an income flow at the long-term rate of about 3 percent—that is, $12,000 per year. The $300,000 in debts costs about $27,000 in interest payments.

The $12,000 pertains only to the first year. After that, it will increase with the rate of inflation. (Any change in the relative or real rental value of land is left out of account; the purpose here is to focus on the problem of a general rise in the price level.) The fact that Mr. A pays a 9 percent rate of interest to buy an asset that yields current returns of 3 percent indicates his expectation that the asset will yield price gains of at least 6 percent.

Suppose that expectations are realized and that prices rise 6 percent each year. How will Mr. A's investment turn out? The expected value of the land at the end of any future period will be the discounted value of the future flow of income plus the discounted value of the land at the end of the period. If expectations are realized, in each successive year the calculations will be the same as in the initial year except that the whole series is increased by 6 percent. Thus, in the first year the value of the land increases by 6 percent, or $24,000.

Pulling together the preceding calculations, Mr. A's results for the initial year are:

Current income from land	$12,000
Increase in value of land	24,000
Cost of interest	–27,000
Total return	$9,000

The $9,000 represents a 9 percent return on Mr. A's $100,000 equity—thus he earns the market rate of interest. The problem that becomes apparent in this example is one of *cash flow*. Since the $24,000 capital gain is unrealized, the net realized return is –$15,000. If Mr. A attempts to repay some of the principal on his loan, the deficit in cash flow will be even larger. Because of the deficit in cash flow, Mr. A will either have to increase his borrowing on the basis of the nominal gain in wealth or else increase his equity by use of income either from labor or from other sources.

Now compare an identical investment by Mr. A *in the absence of inflation*. The interest rate would be 3 percent, there would be no gain in wealth, and the return would be $12,000 – $9,000 or $3,000, a 3 percent rate of return on $100,000 of equity. Again, the net rate of return would be equal to the interest rate, but the problem of cash flow would not arise.

This cash-flow problem is very much like the difficulty that confronts the average citizen in buying a home. Under inflation, both the homeowner and the owner of farm real estate necessarily receive a substantial portion of the returns to their investment in the form of unrealized appreciation in the value of their wealth. But if the investment is financed through borrowing, the interest payments, which reflect anticipated inflation, must be made on a current basis.

The owner of farmland (like the homeowner) will do all right in the end, assuming that the rate of increase in the prices of assets proceeds at the rate implied by the expected-inflation premium in the interest rate. The investment policy of increasing debt in the early years of an investment is not the folly that it might at first appear to be. Indeed, under inflation, one's dollar value of debt must increase if one's value ratio of debts to assets is to remain constant. Nonetheless, inflation creates real difficulties for the expansion or acquisition of farm enterprises by those who rely on debt financing, mainly young farmers. The very serious short-run problem of either going more deeply into debt or of facing an unacceptable squeeze on consumption is undoubtedly the source of many of the complaints that the secretary of agriculture is receiving.

Besides the cash-flow problem, another consequence of inflation is that the indebted farmer is operating in a very risky financial environment. Over and above the usual risks regarding the prices and production of commodities, a change in the general rate of inflation can have a great effect on the results of investment in agriculture. If inflation should accelerate, gains in the price of land will tend to be greater also, and even a 10 percent loan will turn out to be a bargain. But suppose that our government does indeed bite the bullet, thus bringing inflation down. Then the increases in land prices, as in other prices, will tend to decelerate; but the cost of interest will not. The resulting net losses can easily be large.

If interest rates on mortgages should finally fall, the losses could be cut short by refinancing one's debts at the lower interest rates. Long-term interest rates will not adjust downward until the markets become convinced that the long-term rate of inflation has declined. Thus, there is a real risk of substantial, even crippling, losses to leveraged landowners when the rate of inflation is reduced. This is not surprising. Gains from instability in the inflation rate are not a free lunch; if leveraged landholders make extraordinary gains when inflation accelerates, we should expect a coresponding possibility for losses.

There is an obvious solution to these problems: this is for the federal government to pursue macroeconomic policies that will gradually bring inflation under control. This is, of course, not a matter of *agricultural* policy. There is an unfortunate tendency to see current problems as requiring a specifically agricultural, microeconomic approach for their solution. Ironically, in an earlier day when farm interests did see macroeconomic policy as central to their concerns, the main thrust was for *inflationary* policies to solve farmers' low-price problems. This is the central message of William Jennings Bryan's famous "Cross of Gold" speech in 1896. The panacea of that time has the important similarity to those suggested today of calling for the power of the state to solve farmers' problems.

AGRICULTURE AND THE STATE

In addition to the characteristics already discussed, agriculture is distinguished by the atomistic nature of its economic organization. The main decision-making entities are many thousands of individual operators acting independently. The governing of agriculture involves the social institutions in which these actions are undertaken. Most important among the coordinating mechanisms are impersonal and decentralized commodity markets.

Policy issues in the governing of U.S. agriculture may be roughly divided into three categories: the legal framework for transacting in markets, the regulation of market outcomes, and the replacement of the market by more centralized and personalized institutions. Although all these subjects are interesting and important, the main focus of this book is the regulation of market outcomes.

Agriculture has never been immune to influence from politics and the state. Since ancient times it has been a source of tax revenue, in kind if not in money. Sovereigns have long been interested in promoting improvements in production; and direct subsidies to consumers of food, while not an important part of current U.S. commodity policy, exist widely in the world today and go back to the Romans at least. The idea that central political control of agricultural production and prices should be eschewed in favor of a decentralized market mechanism is a quite recent one, having been systematically developed only since the time of Adam Smith.

In practice, a market orientation does not seem ever to have been enthusiastically embraced, even in the United States prior to the New Deal. It was possible for an observer of Washington in the 1920s to say: "One might almost argue that the chief, and perhaps even only aim of legislation is to succor and secure the farmer."[8] Mencken had primarily in mind the protection of farmers from agricultural imports by means of high tariffs. Beyond such measures, the U.S. government's role in agriculture took a dramatic turn in the 1930s, after decades of increasingly insistent demands from farmers that the federal government intervene to increase their receipts from the commodity markets. Exactly why this change occurred when it did is open to question. The Great Depression was undoubtedly the immediate cause; it also provided the proper environment for the changes to take place. In any case, despite some substantial changes, the New Deal approach to agricultural markets, which was adopted in the 1930s, remains the basic approach embodied in current farm programs.

In the period since World War II, new forms of governmental intervention in agriculture and the food sector have developed. U.S. agriculture is today in the process of becoming a regulated industry. The types of regulation include governmental involvement in the pricing of farm products, regulation of farm output and techniques of production, direct governmental provision of services (land and water development; research and extension; storage of products), and regulation of the content, grading, and labeling of food products. Regulation in none of these areas is entirely new, but it has been expanding in scope. This development is not limited to agriculture, but the changes are especially striking there because much of the sector had formerly not been regulated.

Especially notable in recent years is the widening of efforts regarding regulation and redistribution. For example, having a rather full set of programs to protect farmers from low prices, the USDA turns to protecting farmers from losses due to production shortfalls:

> It is our view that modern commercial agriculture must have better protection against natural disaster than is now provided by the hodge-podge of federal disaster programs, limited federal crop insurance, and inadequate private crop insurance. Thus, one of our major legislative proposals in 1979 will be for a comprehensive crop insurance program to adequately protect commercial farmers—allowing them to be efficient in domestic and world markets without undue financial risk.[9]

Moreover, there is a broadening agenda for rural-area programs in general:

> Before long, President Carter will announce the nation's first small town and rural development policy . . . a framework for problem solving, administrative reform and new legislation —all aimed at invigorating rural America.
>
> Let me mention just a few examples of expanded commitment within USDA alone: From fiscal 1975 to the authorization for fiscal 1979, Farmers Home Administration housing loans and grants increased across the board. Individual home loan totals were increased by more than a third, rental housing loans more than doubled, labor housing loans are up by nearly 500 percent and labor housing grants by more than 600 percent.
>
> We have launched two brand new initiatives. One is called the Economic Emergency Loan Program and the other the Limited Resource Loan Program. The first provides some $2 billion through 1980 to help established farmers restructure debt. The

second provides $400 million to help young beginning and limited resource farmers get into farming.

Farmers Home Administration has just guaranteed a $23 million loan for a sunflower seed processing plant in Velva, North Dakota, which will generate electricity as well as produce oil and high-protein animal feed.

A final example—the Home Ownership Assistance Program (HOAP). If Congress appropriates the necessary money, HOAP will make it possible for Farmers Home Administration—for the first time—to reach at least some of the poorest of the rural poor with decent housing.

I think [these programs] add up to a rather remarkable record. I say that because they do, indeed, represent an expanded commitment to provide services, necessities, amenities and opportunities for rural America.[10]

This "expanded commitment" belonged to the Carter administration, which as of this writing is being replaced by a Reagan administration that has been advertised as less interventionist.[11] However, if past experience is a guide, the agenda for governmental involvement in agriculture will not be noticeably smaller under the Republicans than under the Democrats. Indeed, in his first public appearance as designated secretary of agriculture, John Block indicated that he favored intervention in agricultural commodity markets as an explicit tool in foreign policy, a move that goes beyond the reasons for government involvement as formerly stated by either political party. Whether one finds these developments promising or threatening, it is surely proper to take a hard look at the governing of agriculture.

2

Farm-Commodity Policy Today

Two features of the institutional framework for federal farm-commodity policy make economic analysis difficult: (1) the legislation in the area is extremely complex; and (2) wide and often imprecisely defined discretionary authority is given to the executive branch.

The complexity of legislation arises not so much from any particular law but from the way in which many laws fit together. There are about ten important commodity groups—cattle, hogs, poultry, dairy, wheat, feed grains, rice, oilseeds, cotton, tobacco, sugar, peanuts—and at least a hundred minor ones. Congress traditionally, and still today, writes laws on a commodity-by-commodity basis. The legal basis for policy in each of the major areas, and many of the minor ones, is contained in separate laws, written at different times, which have widely varying policy instruments and executive authorities.

Even more confusing is the process by which these laws change over time. Since the 1930s, Congress has enacted an amazing volume of commodity legislation. From first to last, there has been a strong sense in the language of the bills, in the hearings held, and in Congressional committee reports that the programs legislated are to be regarded as experimental. They often have been set up to last only a few years. There is some "permanent" legislation, however, from the 1930s and 1940s, and many later laws are expressed as amendments to it. Thus, our commodity legislation as of 1980 is a hodgepodge of pieces of laws dating from the 1930s to the present, with many provisions lying inactive or never clearly spelled out in practical terms of programs. In consequence, one could read the entire substantial compilation of farm-commodity legislation in law today and at the end have only the foggiest notion of what is actually being done in farm policy.

The current programs for most of the major U.S. agricultural crops were authorized by the Food and Agriculture Act of 1977. That act contains 19 titles, expressed in 132 pages in the statutes.

It is perhaps the most comprehensive piece of farm legislation since the 1930s; it contains a few important innovations, which are based on responses to the upheavals in commodity prices during the early 1970s, experience accumulated under preceding legislation, and changes in the economic and political situation of rural America. While the changes are substantial, they are of an evolutionary rather than a revolutionary nature, reflecting the same basic approach as the New Deal farm programs of the 1930s.

Beyond the good and sufficient political purposes of the Congress, the underlying goals of the 1977 act are so elusive that one might question whether it is worthwhile to consider them. But while the reader should not expect an exposition of the true reasons for congressional behavior, it is useful to have some grasp of the probable intentions of the act.

The text of the legislation is introduced in the statutes as "An Act to provide price and income protection for farmers and assure consumers of an abundance of food and fiber at reasonable prices, and for other purposes" (91 Stat. 913). Without accusing the Congress of being fully disingenuous, there is reason to question whether the stated purpose is a full and balanced statement. In particular, the clause about consumers does not carry the weight that the statement suggests. The real political push for these programs comes from farmers, and it is a push for increased economic well-being. With the decreasing numerical strength of farmers in the voting public, however, consumer and taxpayer interests do have more weight than they had in earlier legislation. The central point is that the law is concerned with distribution of income among interest groups. In this context, each interest group has its own goals, and the overall goals reflected in established legislation are not anyone's goals, but a compromise reached after bargaining and struggle.

INSTRUMENTS OF COMMODITY POLICY

There are many ways in which the economic condition of farmers could be improved, but like previous programs going back to the 1930s, the 1977 act concentrates almost exclusively on increasing the prices that farmers receive for the products they sell. This approach limits considerably the options available for operating the programs. Any program to increase U.S. farm commodity prices has two basic choices: it can attempt to increase the demand for commodities, or it can attempt to reduce the supply. Both approaches have been and continue to be tried.

On the demand side, there have been attempts to boost U.S.

food consumption through various programs of commodity distribution and through food stamps. Successive reforms of the food-stamp program, however, have made it more nearly an income transfer than an in-kind program to subsidize consumption. It is unlikely that this program increases U.S. food consumption enough to provide much economic benefit to farmers. Demand for food is increased by an income transfer only to the extent that spending on food is higher among the recipients than among the taxpayers who finance the transfer. With $8 billion transferred in food stamps and a mean difference of 0.2 between recipients of food stamps and general taxpayers in the proportion of additional income spent on food, there would be a net addition to demand for food of $1.6 billion, or less than 1 percent of the total demand for food.

It is true, however, that producers of some commodities may be materially benefited by food-distribution programs, notably milk in the feeding programs for schools. Even for milk products, distribution through school-lunch programs and other USDA donations amounts only to about 4 percent of U.S. consumption, and this 4 percent is not all a net addition to demand, since some subsidized consumption of milk simply replaces milk consumption that would have occurred anyway.

U.S. programs have been more inventive and persistent in trying to increase demand for U.S. farm products in markets abroad. The USDA maintains an extensive institutional apparatus—in its Foreign Agriculture Service and, more recently, in its Office of the General Sales Manager—for promoting agricultural exports. A particularly useful tool in this effort is credit that is granted to foreign buyers on terms that are more favorable than can usually be obtained through ordinary commercial channels, under the auspices of the Commodity Credit Corporation (CCC).

For some years preceding 1972, direct subsidies for exports had been used to encourage grain exports. A grain exporter could obtain payment for each bushel exported—a sum that was intended to make up the difference between the U.S. price and a lower price on the world market. There were great difficulties, involving millions of dollars, in determining day-to-day movements in the subsidy to be paid per bushel. The program ultimately ran afoul of public opinion in 1972 when subsidies based on low world prices were being paid on exports to the Soviet Union, while an extraordinary surge in demand for grain exports was pushing world-market prices above the U.S. support price. A farmer who sold his wheat in June, 1972, at $1.33 per bushel could have sold the same wheat for $1.75 in

September, $1.90 in October, and $2.00 in November. It was believed that the grain-exporting companies were able to buy low, sell high, and at the same time receive export subsidy payments (until these were ended in September) that amounted to over $100 million. At the same time, the price of wheat that U.S. consumers were paying for in cereal products was increasing.

While both the grain-exporting companies and the farmers were able to claim, though not entirely convincingly, that they did not profit from the episode, it seems clear that U.S. consumer-taxpayers suffered heavy losses. Whether the crafty Soviets, the crafty grain merchants, or the crafty farmers were the principal gainers, the episode still lent itself to expositions with titles like *The Great Grain Robbery*.[1] Explicit export subsidies have not returned as a policy instrument to boost demand.

A more consistently used mechanism for accomplishing the same end is the set of export programs that were established under the Agricultural Trade and Development Act of 1954, which was called "Food for Peace" under the Kennedy Administration and has since become generally known by its number in the Statutes, P.L. 480. P.L. 480 provides a form of permanent food aid, as America's chief weapon in the fight against world hunger. There has been dispute from the beginning as to the effectiveness of P.L. 480 in this regard, mainly because its price-depressing effects in recipient countries tend to discourage production and because the United States tends to cut back the tonnage shipped during the very periods when it is most needed—that is, in years when there is a worldwide food shortage. During these years, prices are highest, so that the costs to taxpayers rise for a given quantity of grain shipped, and the inflation-fighting value of keeping more grain at home becomes apparent.

The political push behind P.L. 480 arises from its role, not in fighting hunger, but in the demand for U.S. farm products. For example, a large amount of P.L. 480 aid to Egypt has for many years taken the form of tobacco. Nonetheless, P.L. 480 cannot be a major factor in supporting U.S. farm prices. In recent years the value of commodities that have been shipped has been in the neighborhood of $1 billion, or roughly 1 percent of the farm value of U.S. agricultural commodities. And while the net addition to demand is probably greater than that resulting from food stamps or school lunches, this effect is reduced by a tendency for P.L. 480 exports to be substituted for ordinary commercial exports.

None of the methods of increasing farm prices by means of

expanding the demand for agricultural commodities has proven sufficient to meet the political demand for higher farm prices. From the earliest programs of the 1930s to those in operation in 1980, the primary policy tool has been to have the government itself provide the demand that is necessary in order to achieve increases in farm prices. This approach in the past has caused prices to be high enough to bring forth costly surpluses of production.

This difficulty leads to the other main approach—namely, reduction in supply. Since the control of individual farmers is very hard to achieve, the most successful mechanism for controlling supply has been through limiting access to marketing channels. But this method has been generally repugnant to farmers, and it has been successful only for tobacco and a few other specialized crops. "Flue-cured" tobacco, the kind that is the principal ingredient in cigarettes, must be marketed through government-approved warehouses. Each farmer has a quota of tobacco marketings—essentially ownership of rights to sell tobacco—which is calculated to put aggregate tobacco sales at a relatively favorable market-clearing price.

For most commodities, supply-control measures have operated by restricting acreage. The acreage-limitation programs that were in effect in 1978 and 1979 worked on the principle of offering producers incentives to "set aside" land that would ordinarily be used for crops. Generally, the set-aside land is to be saved for "soil-conserving" purposes. No cash crops can be raised on it, nor can hay or grass be harvested for forage or grazed as pasture during the principal growing season.

These "set-aside" programs are not mandatory. Farmers do not face criminal charges or severe economic losses if they fail to comply with the programs. Although wheat, corn, barley, and sorghum have set-asides ranging from 10 to 20 percent of planted acreage, as a condition for receiving benefits from price-support and other programs, many farmers have found themselves well able to refuse this offer.

There is even less coercion in the somewhat different "voluntary" programs of acreage diversion that apply to corn, sorghum, and barley. Basically, the U.S. government makes an offer to rent land from farmers, which the farmers may take or leave as they please.

A typology of intervention. The main farm-commodity programs that exist today can be represented as three types of market intervention. The first supports market prices by having the government acquire the excess supply at a policy-determined price. The second approach attains the same market price by controlling supply. The third approach guarantees producers the same price by means of payments that are equal to the difference between the support price and the market price, which is left free to equate supply and demand. Each approach achieves the same price for producers, but the overall effects are quite different (see Appendix for technical details). Under current policy, different commodity programs use different approaches. For some commodities, notably wheat in 1978 and 1979, all three approaches were used simultaneously. An accurate appraisal of 1980 farm policy therefore requires a more detailed discussion of these policy instruments as applied to specific markets.

MARKET SUPPORT PRICES

The most straightforward policy tool for increasing farm incomes through higher commodity prices, at least from a superficial view, is simply to legislate a higher price. Thus, in early 1978 the American Agriculture Movement (AAM) claimed that fairness required a price for wheat of about $5.00 per bushel (compared to market prices at that time in the neighborhood of $2.50), and their view was that the best way to attain the $5.00 price was for Congress simply to pass a law that wheat should not be traded for a price less than $5.00 (actually, the AAM proposal was put forth in terms of "100 percent of parity"). The AAM in 1979 was no longer advocating this simple approach, basically because it had become apparent that such a price would evoke a substantial excess of supply.

Nonetheless, attempts to support prices above market-clearing levels have been a characteristic of U.S. farm-commodity policy since the 1930s. The problem of excess supply is dealt with, in the first instance, by having the government stand ready to buy all that is offered at the support price. In effect, a component is added to market demand, a demand for government stocks which is perfectly elastic at the support price. Hence, the support price is a market-price floor; if supply/demand conditions warrant it, the price may rise above the support price, but the price cannot fall below the floor level.

The institutional mechanism for supporting market prices varies by commodity. For tobacco, the Flue-Cured Tobacco Cooperative Stabilization Corporation, a federally established agency, simply bids at the support prices set for various grades at auctions in the government-assigned warehouses where all flue-cured tobacco must be sold. If the cigarette companies bid higher, they get the tobacco. If they do not, the Stabilization Corporation—that is, the

U.S. taxpayers—buy it. It is then stored and perhaps resold or shipped abroad later under P.L. 480. (Note that the amount of tobacco acquired by the government depends, not only on the overall support price, but on the proper specification of price differentials between different grades and types. The government gets the tobacco that nobody wants, and because of overpricing of lower grades, its purchases have included inordinate quantities of stems and sand, i.e., the worst-looking piles of tobacco leaf in the warehouse.)

The dairy price-support program deserves special notice, because the government here must deal with a perishable commodity. To avoid presenting the taxpayers with ownership of a lake of spoiling milk, price-support operations deal in the market for more readily storable products that are manufactured from milk. Thus, in 1979, Congress passed legislation requiring a price floor at 80 percent of parity (the same type of indexation scheme that was cited with reference to the AAM above), which works out (as of mid 1980) to $12.36 per 100 pounds. USDA analysts then estimate prices for the main manufactured milk products which will be consistent with the $12.36 raw milk price, namely $1.4325 per pound for grade-A butter, $1.325 per pound for grade-A cheddar cheese in 40-pound blocks, and $0.895 per pound for nonfortified nonfat dry milk in 50-pound bags. The milk price is then supported by standing offers of the CCC to buy these products at the announced prices. No one will sell these products for less than can be obtained from the CCC, and no one will manufacture other cheese products, yogurt, ice cream, or whatever, for less return than can be obtained by manufacturing the products that receive direct support. Therefore, the CCC's standing offer supports the prices of *all* milk products and of the milk that goes into them. The CCC stocks are subsequently sold commercially (if prices should happen to rise 5 percent above the support prices) or are sold at a discount or are given away through the domestic or foreign food-aid programs.

The most massive price-support programs historically have been in the grains. The mechanism here is slightly more complex. Producers of grain often desire to store the crop, or part of it, in hopes of postharvest appreciation in the price. But if the harvest-season price is below the support price and if the farmer sells his grain to the government at the support price, he loses his chance to benefit from postharvest appreciation in the price. In order to give the producer the best of both worlds, "nonrecourse loan" programs were developed during the 1930s and continue into the 1980s as the basis for market-price supports for the grains and cotton. To see how the program works, consider wheat. The 1979 "loan rate," the term used for the market support price, was $2.35 per bushel. The producer can put his wheat "under loan," meaning that he places the grain in certified storage on his farm or in commercial facilities, in return for which he receives $2.35 per bushel from the CCC. The money that he receives is a loan from the CCC for a marketing period of less than a year. The CCC loan is in many respects like the loan that one might receive from a bank, using stored grain as collateral. The main difference is the "nonrecourse" feature—if the producer chooses not to repay the loan, the CCC must accept the grain as full payment, and no interest is charged (the producer must pay the storage costs, however). On the other hand, if the market price rises above the loan rate, say to $4.00 (as it did in June, 1979), the producer may redeem the loan by paying back the $2.35, plus an interest rate based on the government's cost of borrowing money. Because this interest rate has tended historically to make CCC loans a cheap source of credit compared to commercial sources, many farmers have put grain under loan even when harvest-season prices are above the loan rate.

In short, the loan program results in an effective season-average price floor in the neighborhood of the loan rate. Farmers redeem the grain if the price is above the loan rate, in which case it goes into normal commercial channels; and the government receives enough grain to hold market price at the loan rate when excess supply exists at that price.

These excess supplies became so large during the 1950s and 1960s that no politically feasible demand-expansion and supply-management schemes could suffice to keep the surplus below embarrassingly large levels. It therefore became clear that support prices would have to be reduced, and in fact, over the 1960s and 1970s, the support prices have been lowered substantially in real terms. Indeed, the historic high loan rate for wheat was the $2.24 price per bushel set in 1954. Adjusted by the Consumer Price Index, the nominal loan rate of $2.50 in 1980 represented more than a 50 percent real decline in market-support prices since 1954.

This drop in support prices, which occurred not only for wheat but for all the major field crops, represents a real increase in the market orientation of U.S. farm programs, but it does not mean that farmers have been abandoned to the rigors of the marketplace. As market-support prices have declined, new policy instruments designed to support farm income without discouraging demand have developed. These instruments try to insure farmers against the economic consequences of low market prices by means of direct payments.

TARGET PRICES / DEFICIENCY PAYMENTS

The Food and Agriculture Act of 1977 continued the evolutionary development of direct payments by linking them to "target prices" based on estimated costs of production. The payments guarantee producers that most of their production costs will be covered, at the same time that they allow markets to clear at a lower price. This basic idea is not new; indeed, it was introduced in the 1930s and was championed by President Truman's secretary of agriculture in the late 1940s, for whom the approach has been labeled the "Brannan Plan." In this approach a target price above the market-clearing level is guaranteed by paying producers the difference between the target price and the market price. Thus, producers know in advance that they will receive (approximately) the target price, but not how much they will receive through the market and how much through government payments. The consequence of the price guarantee is an increased average output, which results in a reduced average market price paid by consumers of the product. Thus, as compared to the market situation in the absence of the program, producers receive more and consumers pay less. This is not a free lunch, because taxpayers must make deficiency payments to the producers which are equal to the difference between the target price and the market price, multiplied by the quantity produced. Payments actually made under this type of program in 1978 were over $1 billion, mostly for wheat.

Although this analysis oversimplifies the programs of the 1977 act, it captures the essential economic impact of them. One complication is that even though the market support prices are lower than they were ten or twenty years ago in real terms, CCC loan programs still exist for the major crops. Consequently, the market price may not be determined by commercial demand but by the loan rate, in which case the deficiency payment is based on the difference between the target price and the loan rate. This puts an upper limit on the rate of payment. Another complication is that a mechanism exists by which payments may be made on only a portion of production, if the secretary of agriculture decides that the aggregate U.S. supply of the crop exceeds "needs." This "allocation factor" may be no lower than 0.8; that is, the producer is only guaranteed payments on 80 percent of his production. A further complication, however, is that by reducing acreage from what was planted in the preceding year, by a percentage that varies from one crop to another and from year to year, a producer can after all be guaranteed an allocation factor of 1.0, or 100 percent.

Still another complication is that payments are based on "program yield," as determined by the government, not on the producer's actual yield per acre in a particular year. Program yield is to be based on a judgment as to each producer's actual normal yield in the past. Thus, if a farmer applies extra fertilizer and increases his output from a given acreage, his payment will not increase until the higher yield is built into the program yield. The result should blunt the farmer's incentive to increase production in response to the target price.

SUPPLY MANAGEMENT

The deficiency-payment approach tends to result in excess supplies being consumed rather than stored, but the simultaneous existence of a loan program still creates a potential for large government-owned stocks. Moreover, while a low loan rate can keep stocks down, there is the problem of deficiency payments becoming larger the greater the difference between target prices and loan rates. Pressure to reduce federal expenditures, which has been increasing since the 1977 act became law, has made the deficiency-payment approach less attractive in Washington. In this situation, the response to large grain supplies in 1977 and 1978 was to return to supply management as a tool for attaining increased farm prices. In this type of program, consumers pay producers directly through higher prices for food and fibers.

The supply-management approach is most fully developed for tobacco, for which the government issues quotas, or rights to market the product, as mentioned above. Less stringent controls exist for peanuts, where production above specified levels must be sold at a reduced price. A few state-level dairy programs and fruit-marketing systems involve production controls. But the most important supply-management issues concern grain acreage.

The acreage-control programs that were in effect in 1978 and 1979 for wheat, corn, barley, and grain sorghum (but were suspended for 1980 crops) were somewhat different for each crop. It is impossible to spell out all the details of these programs, but one may get a feel for their workings by considering a hypothetical example. Suppose that in 1979 Mr. A had 400 acres. If there had been no programs, he would have planted 100 acres of corn, 100 acres of wheat, and 100 acres of soybeans, and he would have had 100 acres of hay and pasture. Suppose the expected market prices were $2.00 per bushel for corn, $2.50 for wheat, and $6.00 for soybeans. The target prices for 1979 crops were $2.20 for corn and $3.40 for wheat, but there was no target price for soybeans. The program incentives in this example would appear to have encouraged a shift to wheat. Mr. A might even have wished to increase his total cropland above the 300-acre level by cutting back on hay or pasture. However, the set-aside programs constrained his choices. There was a 20 percent set-aside on wheat, which meant that if 100 acres of wheat were grown, 20 acres had to be "set aside." The set-aside land is supposed to be essentially idled; it is not to be taken from the 100 acres of hay or pasture nor to be used to add to hay acreage (although enforcement is difficult). The penalty for growing wheat and not setting aside 20 percent is ineligibility for deficiency payments, CCC loans, or the free insurance that is provided against crop failure (which is discussed later) on *any* crop. So Mr. A might not have wished to expand wheat acreage after all. He might even have decided to grow more corn. Corn had a 10 percent set-aside requirement. But there is a big difference in the participation incentives for corn. The $2.20 target price for corn was only 10 percent above that $2.00 expected market price. And $2.00 was the market-price floor, so that $.20 per bushel was the maximum payment benefit for participation. For wheat, the target price was much higher relative to the loan rate, $3.40 compared to $2.35 in 1979. Because the potential deficiency payments were much larger, the incentives to participate in set-asides were greater in wheat.

In order to counter this differential in incentives, the "cross-compliance" provision requires that a farmer cannot receive payments on wheat unless he meets set-aside requirements on corn (and similarly for other protected crops). There remains an incentive problem for corn, however, because contrary to our example, the bulk of U.S. corn is produced on farms that produce a negligible amount of wheat. Therefore, the USDA sweetened the pot available to corn producers. If a farmer had met his 10 percent set-aside requirement, he was eligible to set aside an additional 10 percent of corn acreage in a voluntary diversion program. Participants in this program received a payment of 10 cents per bushel on their normal production of planted acreage. Thus, if Mr. A planted 100 acres of corn, put 10 acres in set-asides and 10 acres in voluntary diversion, and had a normal yield of 100 bushels per acre, he was in full compliance, and he would have received a bonus of 10,000 bu. × 10¢ = $1,000. This bonus is in effect a government rental payment of $100 per acre for the 10 acres that have been idled. Mr. A would probably have found this a good offer.

With an attractive soybean price, Mr. A might well have considered growing more soybeans. This crop has no target-price protection, but on the other hand, it does not involve any set-asides. He might even try putting soybeans on the hay land, so that he could grow corn or wheat on the soybean land and could set aside some of the corn and wheat land. However, the USDA has anticipated such a move. Each farm is assigned a "normal crop acreage"—in Mr. A's case, 300 acres—and the sum of acres planted to all three crops plus set-asides plus voluntary diversions cannot exceed the normal crop acreage. The general result will tend to be that if the soybean price looks favorable, then producers will expand production of soybeans even more than the usual market-oriented response would indicate.

While it is impossible to forecast how producers will respond, in 1978 almost half of feed grain acreage and ⅔ of wheat acreage was enrolled in the program. In 1978, the wheat acreage that was harvested was down 15 percent from 1977, when there were no set-asides, and feed-grain acreage was down 3½ percent. Lower price expectations probably also contributed to the decline in acreage. Yields, however, were up substantially, owing mostly to unprecedented corn-growing weather but also in part because farmers tended to select their poorer land for set-asides. The net result of set-asides was probably about a 4 percent reduction in output from what otherwise would have occurred, although any such estimate is conjectural.

Hard though this may be to believe, the preceding is a considerable oversimplification of the program options and incentives that Mr. A faced. There was a third set of acreage-reduction programs, the "recommended acreage-reduction" programs which influence the allocation factor, mentioned earlier, that determines deficiency-payment coverage. For wheat, there was a wheat-grazing and hay program, which made payments to producers who let cattle graze on wheat instead of harvesting it for grain. For large farms, payment limitations may have come into play which limited payments to any person to $45,000 in 1978 and 1979; $50,000 in 1980. However, this "per person" limit has so many loopholes that it does not seem to be an important factor. The USDA estimates that because of the limitation, 1978 payments were about $24 million ($1\frac{1}{2}$ percent) lower than they would have been without the limitation.

A final set of complications is introduced by the program of disaster payments. This free-insurance scheme was first authorized in the Agriculture and Consumer Protection Act of 1973 and was modified in the 1977 act. Having protected farmers from the vicissitudes of the market by means of target prices, policy makers perhaps naturally turned their attention to what might be done to compensate producers for damage due to crop losses. The program pays producers of wheat, corn, barley, and grain sorghum for any shortfall in yield below 60 percent of their normal yield, times one-half of the target price. Thus, if Mr. A were to suffer a drought and make only 10 bushels per acre, instead of his normal yield of 100, he would receive $(60 - 10) \times \frac{1}{2} \times \$2.20 = \$55$ per acre.

In addition, if a producer is prevented from planting because of "drought, flood, or other natural disaster or condition beyond the control of the producer" (91 Stat. 913, sec. 401), there is a payment on eligible acreage of one-third of the target price, times 75 percent of program yield. Thus, if a wet spring prevents Mr. A from planting 10 acres of his corn land, he will receive 75 bushels per acre, times 73 cents per bushel, or $54.75 per acre. Similar programs provide payments to dairymen and beekeepers in case they suffer various misfortunes, such as a pesticide that renders milk unfit for use or that kills bees. And the Emergency Feed Program pays up to one-half the cost of feed in designated drought areas (in 1978, it paid out $170 million).

More important than any of these programs in terms of dollar volume are emergency disaster loans that are granted in areas designated as disaster areas (due to drought or floods in most cases) and "economic emergency" loans that are granted by the Farmers Home Administration. In 1979, such loans amounted to about $6 billion. However, the net cost to the government is less than for disaster payments, since loans are usually repaid but the payments are not.

The production incentives that result from disaster programs depend on a producer's particular situation, but generally the programs encourage more risky production choices and less care in risk management. This is thought to be an especially serious problem for the beekeepers' indemnity program. Bees are often killed when they are left adjacent to fields or in orchards that are being sprayed with pesticides. The indemnity program seems to have seriously reduced the incentives for beekeepers and sprayers of crops to make arrangements that would avoid these losses. The problems are sufficiently serious that the USDA, which is not noted for objecting to aid to farmers, proposed eliminating from the 1980 budget the appropriations for beekeepers' indemnities. However, the Congressional Appropriations Committees restored the funding for this program.

In crop production, pesticides have an insurance function as well as increasing the mean yield. When pesticides are applied, it is impossible to know the severity of infestation that would have occurred if they had not been used or if a lower dosage had been used. The incentive for use of pesticides is somewhat reduced by the disaster program. More important effects of disaster payments lie in the area of crop selection and in production practices that are followed in marginal production areas, such as areas that are subject to frequent drought or excessive wetness. There are areas in which crop failure is not the exception but the rule, and free insurance encourages overexpansion into these areas. Also, in marginal areas, producers can choose among crops and different varieties of the same crop which vary in their degree of resistance to drought. For example, corn is more susceptible to lack of water than is grain sorghum; but if all goes well, corn is a more profitable crop. The availability of free insurance against crop failure encourages producers in marginal producing areas to make incorrect decisions from the point of view of the efficient allocation of resources. Similarly, in some areas, crop rotations include years when the land lies fallow in order to conserve soil moisture as insurance against lack of rain in the following year. The incentive for this practice is reduced by disaster payments. In general, the program can be characterized as an anticonservation policy.

At the same time, there exists a nominally proconservation policy in the Agriculture Conservation Program (ACP). This program dates back to the 1930s. It provides payments to producers who undertake certain production practices that are thought to promote soil conservation. There is disagreement on exactly what practices should be counted as "soil conserving" and how to set up a program that will promote increased conservation rather than simply subsidize production. The difficulties of this program have been such that in their budgets the last six presidents of the United States, despite wide divergence in the style and substance of their administrations, have all proposed that it be eliminated. But the ACP has had sufficiently well-placed friends on the congressional appropriations committees to survive all challenges to date. In 1978, $240 million in payments were made to farmers under the program.

PROGRAM DETERMINATIONS

The set of production and price-support programs is obviously exceedingly complex. The complexity has existed from the earliest legislation in the area but increased notably in the 1977 act. Indeed, the principal new complications are in the international and stock-management areas, which have not yet been discussed. Not only are the programs complex, but they change substantially and unpredictably from year to year, even within years. Thus, while governmental activity in agriculture reduces many market and production risks that farmers face, it creates a new set of risks and associated opportunities. It seems likely that the considerable entrepreneurial skills that farmers possess have undergone a substantial shift from outsmarting nature and the markets to outsmarting the government.

While the major and most difficult program decisions are made in Washington, the detailed implementation, explanation, and enforcement must be done on a one-on-one basis with farmers. The 2.3 million farmers in the United States are so diverse in enterprise mix, financial organization, economic status, managerial ability, goals, and values, and the farms are so diverse in their physical characteristics, that there are probably no two farmers in the country who find themselves in exactly the same position vis-à-vis the farm programs. For each farmer the following must be determined:

1. The normal crop acreage for all crops.
2. The crops to be counted in normal crop acreage.
3. The program yield for each crop that is eligible for payments.
4. The program acreage for each crop (acreage "planted for harvest"—i.e., not counting grain to be plowed under for nitrogen or grazed and not counting corn to be used as silage).
5. Compliance with requirements for set-aside acreage.
6. Permitted uses of set-aside land.
7. Eligibility for low-yield payments (Are yields below 60 percent of program yields? Were standard practices followed?).
8. Eligibility for prevented-planting payments (e.g., Was a crop planted on a flood plain? Is a failed winter wheat crop going to be planted in the spring?).
9. The suitability of storage for commodities put under CCC loans on farms.
10. Verification of the quality and quantity of commodities stored.
11. Permitted practices and verification for ACP payments.

The institutional apparatus for administering the farm programs is the Agricultural Stabilization and Conservation Service (ASCS). As of 1978, the ASCS had three thousand federal employees and an additional fourteen thousand work years for employees at the county level. The primary policy-implementation instruments of the ASCS are regional, state, and county offices. The county-level organizations, each of which is in charge of three to five committeemen elected by the farmers, are the heart of the system. The state-level officials are appointed by the U.S. secretary of agriculture. The county ASCS committees have broad powers, within guidelines set at the national level, to determine compliance with respect to the items just listed, as well as other determinations for specialized programs. These determinations often make a great deal of financial difference to participating farmers, and the committees are sometimes inconsistent and not always well informed in making their judgments. For example, consider this dicussion from hearings on the 1980 ASCS budget:

> Mr. Andrews: . . . The Director [of a county ASCS office] sent out postcards that read as follows: "You know that low area that you couldn't plow or plant? It may be used as set aside if it contains at least two acres. You know that field you are planning to tile? Why not designate that as an area for the set aside. Sign up now. Be safe, not sorry. You may drop out later with no penalties." Now, this came from an Illinois reader. Did your ASCS office in fact send out this type of material?
>
> Mr. Fitzgerald: There are 2,734 offices, and some offices might have sent out something like that.[2]

The federal regulations on this matter specify that the following lands are not eligible for set-asides: "Turn rows, drainage ditches, wet low-lying areas, droughty knobs or banks, areas rejected by the county committee because of their small size or shape. . . ." The regulations leave plenty of leeway in which the county committees may operate, and the local pressures are always to be lenient. This creates difficulties for set-asides, in that producers want to divert land that is unproductive. The problems may be even greater for obtaining approval of ACP cost-sharing projects, for establishing program yields, and for filing claims for disaster payments.

The range of activities is illustrated by data on the largest and smallest payments made by the ASCS in fiscal 1978 under a program that I have not discussed up to this point, the Emergency Conservation Program (which does *not* deal with encouraging conservation practices where there is imminent danger of environmental degradation, as the title might suggest). The largest payment was $87,242 to Dixie Ranches of El Centro, California, for "removing debris from farmland," "grading, shaping, releveling, or similar measures," and "restoring structures and other installations."[3] The smallest was $86 to a farmer in Massachusetts for restoring fences.

The combinations of programs can lead to quite bizarre results. Under earlier long-term retirement programs, conservation-program funding could be used to build tiling systems to carry irrigation water from wells on acreage that had been diverted under production-control programs, though the water was to be used to increase yields on adjacent land that had been kept in production. The disaster-payments programs for feed grains acquire a special irony when considered in conjunction with the Emergency Feed Program that is available to farmers in some areas. Using 1978 data, consider a farmer who for $1.90 per bushel can produce corn which he normally feeds to livestock on his farm. He suffers a total loss of his crop, in which he has invested $1.40 per bushel in pre-harvest costs (based on normal yield). Then he has to buy corn for $2.00 to obtain feed. Without any program, his cost for corn that he feeds is $3.40 per bushel in the disaster year. But with the programs, he gets a disaster payment of $.63 per bushel (one-half of the target price of $2.10 on 60 percent of the production that he has lost), plus a rebate on the purchased corn of $1.00, for total payments of $1.63. Subtracting $1.63 from $3.40 leaves $1.77 as his cost for grain fed. The farmer is better off having his crop wiped out than he is in a normal year when his feed costs him $1.90 per bushel to produce!

For prevented plantings, the situation is even more extreme. The farmer may only have lost his fixed costs of land ownership, say $.50 per bushel (based on normal yield), so that with a price of $2.00 for purchased corn, he would pay $2.50 for feed with no programs. But he gets a prevented-planting payment of one-third the target price on 75 percent of normal yield—that is, $.75 \times 1/3 \times \$2.10 = \$.52$—and a $1.00 rebate on feed purchases, so the cost of feed under disaster is $\$2.50 - \$1.52 = \$.98$ per bushel, roughly half of what it would cost to grow his own feed in a normal year.

The secretary of agriculture may not permit the program to be administered in such a way as to cause such anomalies, even though the legislation invites him to do so, and the system of having county committees invites abuse. The subject is brought up as an example of how the many small and independent programs contained in the 1977 act may combine to produce strange results in the existing institutional environment.

The cost of programs. In chapter 4, I will discuss in detail the overall effects of these programs as well as others yet to be mentioned. It may be helpful at this stage to consolidate some information about the level of governmental activity that was associated with the various programs in the budget for fiscal year 1978. These data are shown in table 2.1. These figures include by no means all the programs of the USDA, and they do not include programs that benefit farmers which are administered by other agencies—for example, loans made by the Small Business Administration. My

TABLE 2.1

SELECTED COSTS ASSOCIATED WITH U.S. FARM PROGRAMS, FISCAL YEAR 1978

PROGRAM		COST (in millions)
CCC Operations, Including P.L. 480		
Storage, handling, transportation expense		$ 62
Loans written off		9
Producer storage payments		413
Net loss on sales and donation of commodities		234
P.L. 480 food aid (title II)*		328
P.L. 480 ocean freight		193
Administrative expense		46
Commodity Programs		
Price guarantee payments:	wheat	996
	barley, sorghum, rice	228
	sugar	212
	wool	36
Payments for voluntarily idled land:	corn	162
	barley, sorghum, wheat, cotton	27
Disaster payments: grains, cotton, rice		526
Other Programs		
Agricultural Conservation Program		226
Emergency Feed Program		169
Emergency Conservation Program		30
Forestry Incentive and Water Bank Program		23
ASCS Salaries and Expenses		329
	TOTAL	$4,249

SOURCE: United States, House of Representatives, *Agricultural, Rural Development and Related Agencies Appropriations for 1980,* Hearings, pt. 5, Agricultural Programs.

* This figure does not include $671 million sold for credit and foreign currency under title I, P.L. 480. On the order of half this sum should eventually be recovered, so it is not all net cost.

purpose is mainly to show how the cost of many small programs can add up.

COMMODITY RESERVES AND INTERNATIONAL AGRICULTURAL POLICY

"The Farmer-Owned Reserve Program has become the cornerstone of [our] food and agriculture policy," according to Secretary Bergland.[4]

The reader may feel some despair at having gone through a description of an edifice that is as complex as the one sketched out in the preceding pages, only to find that the cornerstone has been left out. Actually, the type of programs that have been outlined so far are historically and currently the foundations of U.S. farm policy. It is true, however, that programs under the 1977 act give increased prominence to efforts to stabilize prices of internationally important farm products, notably grains, and to international trade problems generally.

Grain-storage programs. The idea of using grain that has been acquired by the CCC for price stabilization instead of surplus disposal—by releasing stocks in high-price years intentionally to hold prices down instead of trying to send stocks into domestic or foreign uses that would not reduce commercial demand—is not new. The approach has often been given at least lip service in phrases such as "ever-normal granary" which have been attached to government stocks.

Past experience with government commodity-storage programs has revealed several problem areas. First, there is extensive storage of commodities in the private sector, which is sensitive to government storage regimes. For easily storable commodities like grains, a large amount of such storage consists of farmers holding from the market the grain that they have produced. For commodities that are less easily stored, commercial storage predominates; for example, in cold storage of dairy products and meats. Moreover, private-sector storage occurs on farms by methods that are not apparent as such at first glance. Ranchers and cattle feeders can effectively store meat on the hoof by varying their handling of cattle. In the past, New England producers stored grain in the form of whiskey. Today, to carry the cattle example one step backwards, cattle can function as stored-up grain or forage—in a sense, this is what they are.

Government storage can result in substantial reductions in the incentive for private storage. For example, the CCC has at times followed a policy of buying dairy products at the support price and then reselling them at 1.05 times the support price. The difference between the purchase price and the sale price does not cover the cost of even half a year's interest on funds that are tied up in commodities, not to mention physical storage costs. Thus, the commercial trade tends to be content to let the CCC do the industry's precautionary storage.

Another problem with government stockpiling of commodities is that stock-release policy is potentially sensitive and becomes difficult to conduct on the basis of economic criteria. Producers tend to be wary of stock release as a depressant on their incomes. When prices have risen only a little, they think that the time is not ripe for release. When prices have risen a lot, they think that the CCC is destroying their opportunity to make a little money after all those years of losses. Even when explicit CCC rules regarding release have been spelled out in advance, they have been politically adjustable and hence unreliable guides for the private trade's storage.

The programs under the 1977 act attempt to avoid these difficulties for grain by relying on a "farmer-owned reserve" for grains. This is a subsidy program for on-farm storage. The most important means by which grain is placed in the reserve program is "extended loan reseal." Suppose a producer harvests his crop and puts it under a nine-months CCC loan. The extended-loan feature permits the farmer, when the nine months are up, another option besides delivery to the CCC or redeeming the loan (and paying the interest on it). He can retain the grain in certified on-farm or commercial storage. In order to encourage participation, the farmer receives annually an advance payment, $.25 per bushel for wheat in 1979, which is intended to cover storage costs. As of 1979, the extended loans cost the producer 6 percent interest for the first year, with no interest thereafter. The loans are offered for a three-year period, and the subsidies are contingent on a farmer's agreeing to hold the grain for three years. These features are subject to change at the discretion of the secretary of agriculture.

An important new feature of the 1977 act is the mechanism for releasing stored grain to the market. The government retains the option of ending the subsidies and calling in the loans if market prices should rise. For example, when the market price of wheat reaches 140 percent of the loan rate (e.g., in 1979 if market price had risen to $\$2.35 \times 1.4 = \$3.29$), the producer could have redeemed the loan without penalty and sold the grain. But if the loan is redeemed when the price is less than 140 percent of the loan rate, the farmer must repay all storage costs. The farmer must pay at least

140 percent of the loan rate on each bushel that he redeems. In 1979, when the market price reached 175 percent of the loan rate ($4.11), the loans would have been called, i.e., the producer would have been required to redeem the loan.

The idea is that when prices become this high, wheat should be forced onto the market and should no longer be stockpiled. However, the CCC's ability to call the loan is not a call option on the grain itself, and the producer may keep it in storage under self- or commercial finance. As of mid 1980, it appeared that in the neighborhood of 1,200 million bushels of wheat and corn stocks—an amount equal to almost half the total carryover of grain—would be in this program.

An additional element of reserve stocks is a government-held International Emergency Wheat Reserve of up to 220 million bushels. Grain delivered to the CCC under the loan program can be placed in this reserve. Release of this or other CCC grain cannot occur when the farm price is less than 150 percent of the loan rate so long as farmer-held reserves exist.

In addition to these programs, there is a farm-storage-facility loan program to subsidize the building of on-farm storage capacity. The program provides up to $100,000 and up to 85 percent of the cost of the facilities on relatively favorable terms. During 1978, under this program, storage capacity for 750 million bushels was constructed at a cost of $650 million.

It is important to note the particular dates attached to almost every program detail in the preceding paragraphs. Almost every policy parameter in the reserve programs was changed in early 1980 in order to cope with the embargo of shipments to the Soviet Union. As a stabilization program, it is disquietingly variable.

International aspects of farm programs. For the grains, international considerations are central to the grain-reserve programs as well as to the price-support activities. The United States from 1977 to 1979 was an active participant in negotiations for an international wheat agreement, in which the major wheat exporting and importing countries would engage in joint price-stabilization measures, primarily by means of buffer stocks. The U.S. storage programs were designed, in part, to fit in with an internationally coordinated approach. However, the negotiations broke down in 1979, and it appears that the United States will have to conduct any price-stabilization efforts unilaterally.

World markets are important because so much of the U.S. crops are exported. About two-thirds of 1978 U.S. wheat production was exported, along with an estimated 30 percent of corn, 42 percent of soybeans, 55 percent of cotton, and 58 percent of rice. The quantities are large enough to have a significant impact in the world markets, especially for feed grains and soybeans. Even for rice, a relatively minor U.S. crop, the production of which is about one-eighth of U.S. wheat production and one-thirtieth of its corn production, the United States is the world's largest exporter.

While U.S. exports are important determinants of world market prices, the demand function for U.S. exports is still fairly elastic. Consequently, if U.S. support prices rise, demand declines much more than one might expect if one were to consider only the normally inelastic demand for food. For U.S. wheat, for example, recent research indicates that the total elasticity of demand—including domestic use, exports, and speculative stocks—may be in the neighborhood of –1.0. That is, if U.S. wheat supplies fall by 10 percent, the price can be expected to rise by about 10 percent. And if effective market-support prices are raised 10 percent, government stocks will increase by 10 to 15 percent of supply, or around 300 to 400 million bushels. (This calculation ignores supply response to the market-support price, since under current programs, supply incentives are determined by the target price when market prices are low.)

This order of magnitude for unwanted stocks would be quite costly. At $.25 per bushel for annual storage costs and at a 10 percent interest rate, the annual cost of storing 400 million bushels of wheat is about $200 million. In an attempt to avoid such possibilities, the 1977 act gives the secretary of agriculture authority to reduce the loan rate in the year following a year when the average market price received by farmers is within 105 percent of the loan rate. This authority has not yet been exercised, however, even though it could have been for wheat in 1978, and it is questionable whether the political will can be mustered to do so. The pressure seems to be too great in favor of production controls rather than price cuts when excess supplies appear.

Production controls, too, take on a new aspect in the context of the world market. When U.S. production is reduced, this supports not only the U.S. price but also the world price. To see the problem that is involved, suppose one desired to increase the price of wheat by 10 percent. If only U.S. domestic demand were involved, with an elasticity of, say –.2, then a 2 percent reduction in output via set-asides would suffice. But if the demand for exports should make the total elasticity of demand –1.5, then a 15 percent reduction in output would be necessary. The extra 13 percent (roughly 4 million metric tons) is the quantity of wheat that must be withdrawn from the world wheat market in order to get the world price increased 10 percent so that it will be consistent with the U.S. price.

In the world context, what is done with grain set-asides is equivalent to what the OPEC nations do by controlling the output of oil. The United States is in fact attempting to be a one-country wheat cartel. Realization of the difficulty and costliness of this attempt has led farm groups to urge that an OWEC be formed of the principal wheat exporters—the United States, Canada, Australia, and Argentina. Indeed USDA officials have had many discussions, with the Canadians especially, about coordinating grain exports. These efforts have to date come to naught, as well they might, since the United States has shown itself ready to engage in costly unilateral action to support world prices. Grain-producing interests in these countries might well content themselves with being free riders on U.S. price-support efforts.

Being tied to a world-market price puts severe constraints on U.S. farm policy. For other countries, almost all of which have much less impact on world prices than does the United States, there is even less leeway. They are essentially price takers in world markets. The almost-universal response to this state of affairs is to try to separate internal commodity prices from world prices by means of duties or quotas on the import side and by means of taxes, subsidies, or quantitative controls on the export side. Thus, while grain prices in internal U.S. markets move parallel to world market prices, this is not the case for most other countries. Prices received by farmers and paid by consumers are as much an artifact of different countries' policies as of supply/demand conditions in these countries.

U.S. agricultural export policy is not totally market oriented. In addition to promoting exports under P.L. 480 and other means, the target-price/deficiency-payment approach can be viewed as a de facto export subsidy. The target price induces an expansion of output, which clears the market at a reduced price. Consequently, the target price has the effect of increasing the demand for U.S. exports and of reducing world prices just as an export subsidy would. The difference is that an export subsidy creates an artificial divergence between the U.S. price and the world price, while the target price does not. Note, however, that since the target price is based on the estimated costs of production, the market price at which grain is exported must be on average below cost, so that according to our own criteria, we are "dumping" grain on world markets.

Two farm commodities for which current U.S. price-support programs involve import restrictions are milk and sugar. Imports of dairy products, particularly cheese, are constrained by quotas. In the absence of quotas, these imports would be substantially larger, and prices for U.S. dairy products would be lower. This effect would occur for simple commercial reasons for the specialty cheeses of various countries, and it would tend to occur, perhaps in more massive quantities, for butter and powdered milk as an outlet for surpluses created in other countries by their own price-support programs. The European Economic Community (EEC) is especially noted for its "butter mountain" that accumulates periodically and is sold, for example, to the Soviets for around 10 percent of its cost to the EEC states, and for its stock of powdered milk, which it sells at a loss to EEC farmers who then feed it to calves (which then grow up to produce more milk).

The sugar crops (sugar beets and sugar cane) have been among the most heavily protected of all U.S. farm commodities. Import policies held the U.S. market price for raw sugar at almost twice the world (offshore) market level in 1978/79. This is especially striking in that there are only about fifteen thousand producers of sugar cane and sugar beets in the United States (compared to more than a million producers of corn, for example). Nonetheless, sugar is important to consumers; about 30 percent of the U.S. caloric intake comes from this product.

The sugar program under the 1977 act deserves special mention because it is an almost entirely new program, which still is in a state of flux, and because sugar is the most important domestically produced farm commodity for which imports account for a substantial fraction of U.S. consumption. In 1978, a target price of 15 cents per pound of raw sugar, New York basis, was established, which was about double the world price (offshore at New York). However, deficiency payments are not the main mechanism that is used to make up the difference. The program retained the 1.875 cent per pound tariff that was established when the Ford administration tripled the tariff just prior to the 1976 elections and added "import fees," which have fluctuated in the range of 3 cents per pound. The 1977 act also authorized import quotas for sugar, which could be used to achieve any desired target price, as well as a loan/purchase program. Thus, the secretary of agriculture has at his disposal four tools—deficiency payments, the tariff (including import fees under Section 22 of the Agricultural Adjustment Act of 1933), the quota, and the loan/purchase program—any combination of which could be used in establishing a support price.

The 1978 and 1979 programs involved payments to producers, CCC loans, and import tariffs and fees. As a result, the CCC acquired about 200,000 tons of sugar, which is costly to store and is not easy to dispose of through the usual channels for donating CCC stocks. By 1980, the world sugar market had tightened, with market prices being over 30 cents per pound, and tariff and import fees on sugar had been reduced to the statutory minimum of 0.625 cents per pound. The sugar programs were essentially put on hold.

With respect to imports of food products that are not produced in this country, principally tropical products, the United States has become increasingly sympathetic to international agreements, following its longstanding support of and membership in the International Coffee Agreement. These, which are intended mainly to provide aid internationally for commodity producers in low-income countries, are somewhat analogous to the aid that is provided for U.S. producers by U.S. farm programs. However, the consuming countries have argued that such agreements should be purely stabilization arrangements and not price-support programs such as the ones that are used in U.S. markets. This has led to internationally coordinated buffer stocks with prescribed buying and selling prices. Nonetheless, it appears that the international agreements that have been negotiated to date—for coffee, cocoa, and sugar—will be more effective in putting a floor rather than a ceiling on the market price, although the history of commodity agreements suggests that they will not be very effective at either.[5]

The International Sugar Agreement (ISA) provides the latest test case. After international negotiations and domestic political discussions that lasted more than three years, the United States formally became a full participant in the ISA in 1980. Although the ISA contains buffer-stock arrangements, its principal economic tool appears to consist of provisions by which sugar exporters will jointly restrain exports when the price falls to a floor trigger. Thus it is a cyclically triggered international cartel, which could more suitably be named OSEC, except that net consuming nations like the United States have chosen to abet this effort. This is supposed to be justified as a pursuit of market stabilization.

Imports of fresh vegetables from Mexico and Central America during the winter months, when they cannot be grown in most of the United States, would seem to be a natural commercial venture yielding the classic benefits of trade. However, the U.S. winter vegetable-growing industry, especially the Florida tomato growers, have been attempting to limit these imports. They argue that Mexican tomato growers export below cost and that the fact that the Mexican growers are exempt from U.S. restrictions on the use of pesticides puts the U.S. industry at an unfair disadvantage.[6] However, as of mid 1980, the tomato growers have been unable to obtain relief, except for restrictions on packing procedures, despite appeals to the Treasury Department (under anti-dumping statutes), the Congress, and the U.S. International Trade Commission.

The idea of putting restraints on imports in order to support farm income is not new. It antedates the Depression-era programs, whose descendants we have been discussing.

The principal remaining use of U.S. import restrictions, in the absence of any other price-support program, are quantitative import restrictions on some meats and a long-standing 3-cent-per-pound tariff on beef. The Meat Import Act of 1964, with a formula revised by new legislation in 1979, established upper limits on imports of fresh, chilled, or frozen beef (as well as goat meat and mutton). The quantitative restrictions work by means of "voluntary restraint agreements" by which beef-exporting countries (principally Austrialia and New Zealand) agree to limit the amount that they will ship to the United States each year. Countries agree to these restraints because if they do not, they will be limited still further by means of quotas.

Restrictions on exports constitute one of the few U.S. policy interventions apparently undertaken with the intention of aiding food consumers at the expense of producers.[7] The U.S. Constitution (Article 1, Section 9) explicitly forbids the taxation of exports, but quantitative restrictions can be and have been used to the same end. However, the only actions of this sort that have been undertaken were temporary embargoes on soybean exports in 1973 and on grain sales to the Soviet Union and, for a short time, to Eastern Europe, in 1974 and 1975. It is unlikely that these restrictions had any effect on the supply-demand balance either domestically or abroad, but they did have repercussions that still have influence. First, foreign purchasers of U.S. products naturally found it prudent to increase the diversity of their sources for imports. Second, the political reaction from producers was so strong that it induced both presidential candidates in 1976 to make very strong promises to avoid future restraints on exports. In 1980, however, grain sales were again halted to the Soviet Union, this time as a weapon in a diplomatic struggle. It seems inevitable that farmers will henceforth be wary of the export market as an outlet for their products, just as foreign importers of grain will be wary of the United States as a source of supply.

Notwithstanding the interventions just outlined, the United States was a strong supporter of liberalized world trade in agricultural products in the Multilateral Trade Negotiations (MTNs) concluded in Geneva in 1979. The United States in fiscal 1980 exported $40 billion in agricultural products (roughly 30 percent of the total value of farm production) while importing $18 billion in agricultural products.[8] Our major trading partners have substantially greater degrees of protection than we do for agricultural commodities, and the overall interests of U.S. agriculture, as well as the country as a whole, must surely lie in the general liberalization of trade. While steps were taken in the trade agreements reached at the MTNs which will increase the trade flows of agricultural commodities, they are relatively minor, and they work within existing protectionist structures rather than providing any fundamental shift towards a more trade-oriented world agricultural economy.

AN OVERVIEW OF U.S. GOVERNMENT INTERVENTION IN FARM-COMMODITY MARKETS

The discussion of U.S. commodity programs in this chapter undoubtedly gives the impression of a hopelessly complicated set of unrelated interventions. This impression could be magnified by adding discussion of the programs for peanuts, extra-long-staple cotton, burley tobacco, rice, mohair, and honey, all of which differ from the programs discussed above. Even for commodities that have no explicitly legislated program, ad hoc interventions can be undertaken under general authorities granted to the secretary of agriculture. For example, in early 1979 a program was undertaken to buy potatoes and to have them used for cattle feed in order to increase a depressed price for this commodity, owing to an exceptionally large 1978 harvest. There is no explicit target price, support price, or legislative statement of purpose to guide one's understanding of this program.

Nonetheless, the 1977 act represents a significant attempt to attain better policy coordination among the various crops than had existed in the past. And there has been a general thrust toward market orientation. That is, while payments to producers have increased in importance and while supply-management initiatives persist, less attempt is being made to hold market prices at levels that are not justified by underlying conditions of supply and demand. Moreover, many of the most important farm-commodity markets are not subject to any significant systematic intervention. This is true of soybeans and other oilseed crops, most fruits and vegetables, and forage crops for livestock. Most important, hogs, broilers, and the cattle business (apart from the relatively minor import restraints) do not have programs to regulate prices or production. These and other unregulated commodities accounted for $52 billion of the $103 billion in 1977 farm production.

3

The Regulation of Marketing and the Middleman

Statements about food policy by public figures abound in sympathy for all who produce and consume, but it is difficult to find a good word about the middleman. Middlemen have traditionally been depicted as irresponsible fellows who trick farmers into selling farm products at low prices, mix those products with various worthless but cheap ingredients, and then, through deceit and fraud, persuade consumers to buy food products at exorbitant prices. Included in this mistrusted company are also the bankers who supply farmers with credit; the businesses that supply machinery, pesticides, and other goods and services used in agricultural production; and sometimes, absentee landlords who own land that is cultivated by tenant farmers.

The most important distinguishing feature of middlemen, which contributes to their disrepute, is the ability to exercise market power in a way that farmers or consumers cannot attain. In the market for a food product there are typically thousands of final consumers and farmers who have to deal with a handful of middlemen. Beef cattle in Kansas, for example, go from hundreds of feedlots to a small number of slaughtering plants, and are then shipped by one of the relatively few transportation enterprises through a few retail outlets in each town or city to thousands of consumers in each city.

The degree of market power that is possessed by businesses at various points in the marketing channel and how much is gained and by whom as a result of exercising such power are matters of long debate. It is a political fact of life that the farmers' feelings of powerlessness and exploitation at the hands of middlemen have opened the way for governmental actions in the food and fiber sector of agriculture, apart from regulation of the commodity markets.

Much of this intervention is aimed at regulating commercial enterprises in the food industry. It is not a new area of governmental action. Federal food-safety and health legislation went into effect under the Food and Drug Act, enacted in 1906. The Packers and Stockyards Administration has been in operation since the 1920s. However, more important than the regulation of private business has been the establishment of new public and quasi-public enterprises designed to supplement the middlemen or remove the need for them.

In order to understand agricultural policy, it is important to acquire some appreciation of the characteristics of these institutions that have been created to aid farmers in marketing their products and in purchasing their inputs. A few, like the Tennessee Valley Authority, are basically federally owned utilities. Others, like the Rural Electrification Administration, are basically conduits to channel governmentally raised capital into politically favored lines of investment. Some, notably the Federal Land Banks, Federal Intermediate Credit Banks, and Production Credit Associations, which were set up to provide long- and short-term credit to farmers, were established with public capital but have since paid it back and today function largely as private institutions. The Federal Crop Insurance Corporation (FCIC) exists alongside the Disaster Payments Program to sell crop insurance to farmers on a wide variety of crops. The Farmers Home Administration (FHA) is a government agency within the USDA that provides loans for many of the same purposes as the independent farm-credit associations, but it accepts riskier and broader social-purpose loans to rural residents.

From the farmers' point of view, the most important institution-building policies have been legislation that permits farmers to form voluntary associations which, they hope, will increase their market power as opposed to the industries that sell inputs to them and ones that buy their output from them. Two institutional creations are central to the governmental effort to increase the farmer's bargaining power: cooperatives and marketing orders. Both institutions are complex, and there is much disagreement even among experts on some of their legal and economic characteristics.

COOPERATIVES

The underlying idea of a cooperative is that it is a way for a voluntary association of individuals to avoid the necessity of dealing with profit-seeking business organizations. Thus, a group of consumers may band together to distribute staple food items, thus enabling them to avoid retail food outlets. A group of cattle feeders may jointly purchase and distribute cattle feed, selling "at cost," to avoid commercial millers. A group of vegetable growers may pool and jointly market their products in order to bypass commercial wholesalers.

This underlying idea is quickly weakened when a cooperative attains any significant scale of operation. It must then have full-time employees, a managerial team, and its own financing. Given these features, it can no longer function at the direction of all members jointly but requires a representative system of direction through a board of directors. In short, it is no longer a joint voluntary activity carried out by individuals, but is a business in itself. Indeed, it is functionally almost indistinguishable from a profit-seeking corporation, which, after all, is also a voluntary association of individuals. The main functional difference is that the principal claimants on a cooperative's earnings are its customers, while this is not so for corporations. Thus, a cooperative's earnings that are not retained in the business are distributed primarily as "patronage refunds" instead of as dividends (although some cooperatives do issue stock and pay dividends in addition).

An important reason for the survival and expansion of cooperatives in agriculture appears to be the tax treatment accorded to patronage refunds. They are not counted as income to the cooperative and are therefore not taxed (although they are taxed as personal income to the recipient). Moreover, earnings that are retained in the business, after they have been allocated to members' accounts for later payment, are not taxed as far as the co-op is concerned (although these retained funds, too, are taxable income for the members). This may seem to represent a minor tax difference, but it is not. Compare the following data for a corporation and a cooperative:

	Corporation	*Cooperative*
Sales	$5,000,000	$5,000,000
Costs (depreciation, etc.)	4,500,000	4,500,000
Income	500,000	500,000
Dividends (patronage refunds)	200,000	200,000
Corporate taxes (48 %)	240,000	0
Retained in the business	60,000	300,000
Personal taxes (25% rate)	50,000	125,000
Owners' income after taxes	210,000	375,000
Owners' Equity	$3,000,000	$3,000,000

The corporation returns $210,000 to its shareholders, a posttax rate of return of 7 percent on their equity. The cooperative returns

$375,000, for a posttax rate of return of 13 percent. This illustration suggests why the cooperative form of business may be relatively attractive. Indeed, one would expect that if managerial capacity were equivalent, cooperatives could offer so much better terms to their members than could corporations to their customers that ordinary profit-seeking businesses would be driven out of the farm-supply industry. There is a potential for problems in raising debt capital, if managerial problems are pervasive and cooperatives tend to get into commercially suspect enterprises. However, difficulties here may be offset by the federally chartered Bank for Cooperatives, which the cooperatives themselves own and which accounted for 64 percent of the outstanding debt of cooperatives in 1970.[1] Under these financial umbrellas, cooperatives have flourished in the U.S. food system, spreading into many areas in the purchase of farm inputs and the sale of farm products. In 1976 there were over twenty-seven hundred farm-supply cooperatives, with memberships of three million. They accounted for about one-fourth of total farm expenditures on purchased inputs.

While service-providing cooperatives raise questions of fairness in taxation, more far-reaching policy issues involve the use of cooperatives as associations for the purpose of collective bargaining by farmers. In this role, they need not actually buy or sell any products, but can simply act as a bargaining agent. For example, dairy cooperatives need not replace milk handlers; instead, they may concentrate on reaching collective terms regarding the sale of milk to profit-seeking handlers. This function was substantially enhanced by the Capper-Volstead Act of 1922, which established many of the basic rights and obligations of cooperatives in agriculture.

The Capper-Volstead Act attempts to increase the bargaining power of farmers by shielding farmer cooperatives from antitrust laws. Nonetheless, cooperatives do not have the legal right to monopolize trade that is sometimes attributed to them. Certain collective-marketing practices are exempted from antitrust laws, but "undue price enhancement" is expressly prohibited. However, enforcement is entrusted, not to the Department of Justice, but to the secretary of agriculture. In seventy-five years no secretary has found an instance of undue price enhancement, and it is doubtful that there has been a serious search for this phenomenon. In recent years, investigators from the Justice Department, the Federal Trade Commission, the General Accounting Office (GAO), and the Executive Office of the President have brought forth evidence that there ought to be more serious attention paid to the issue.

Actually, there are good reasons to doubt that cooperatives in and of themselves could raise prices very much, because of the difficulties of controlling the output of members, new entrants, and most importantly free riders who remain outside the cooperative but can sell at the market prices established by the cooperative. What the promoters of farmer bargaining power are ultimately seeking is a cartel—a means by which independent producing agents may jointly restrain marketings and thereby attain effective monopolistic power. The problems inherent in the organization and maintenance of cartels are such as to make the U.S. government's permission to form them in agriculture an empty gesture. Practical proof of this proposition was provided many times by failures, especially during the 1920s.

MARKETING ORDERS

Effective cartels can exist only under governmental management, with the state's power to punish the offenders of marketing restraints. Given governmental sanction, a cartel of very many members becomes a plausible proposition. Indeed, the flue-cured tobacco program that was discussed earlier is a fine example of a cartel, even though it contains many thousands of independent producers. The government, as manager of the cartel, each year sets a production goal, allocates it among producers, enforces a limit on sales of each member, controls entry absolutely, and allows no one to buy or sell the product outside the cartel's channels. OPEC should have such clout over its members! The price supports, "four-leaf" program, and foreign-aid shipments of tobacco are just frosting on the cake.

Farm interests attained the governmental tools that they needed in order to make cooperatives more truly effective bargaining agencies through legislation in the 1930s, which culminated in the Agricultural Marketing Agreement Act of 1937. This legislation developed the institution of federal marketing orders. A marketing order is a contract among all producers and handlers of a specified commodity in a particular market area. They are established by the Agricultural Marketing Service of the USDA, if, after rather involved investigatory procedures, it is determined that producers in a market desire one. There is no narrowly prescribed legal procedure for initiating the consideration of a marketing order or for determining if the producers desire one. After an order is provisionally drawn up by the USDA, producers vote, and if a majority, or in some cases two-thirds or three-fourths, favor the order, it goes into effect. Handlers cannot veto it. Cooperatives often are the moving force for the creation or modification of a marketing order: for example, by merger of separate orders into one larger market-order area. When an order is established, it is binding on all producers and handlers. The order may deal with pricing or, as is more typical, only with grades, standards, or regulating seasonal flows to market.

The administrative costs for operating federal marketing orders are paid out of assessments on handlers and, in the normal course of events, are incorporated in the consumer price of retail food to the same extent as are other costs of doing business.

In 1978, there were in force some ninety-four marketing orders, forty-seven regional milk markets and forty-seven for various fruit and vegetable crops—for example, oranges, grapefruit, lemons, olives, peaches, avocados, prunes, cranberries, celery, onions, potatoes, almonds, walnuts, filberts, and hops. These orders accounted for $13 billion in sales of agricultural products, 12 percent of all farm cash receipts. There are in addition some state-level orders, notably in California.

Marketing orders are a natural adjunct to marketing or bargaining cooperatives. They obviate the free-rider problem, because the government enforces the terms of the orders. The secretary of agriculture has wide authority in establishing, disbanding, regulating, and enforcing the provisions of an order. Nonetheless, almost all of these marketing orders lack the supply-control features that are necessary in order to make them cartels in the classical sense. For some commodities—for example, potatoes—there are several orders that, taken together, still omit most U.S. production. Many other orders lack price-influencing activities altogether. So one should not overstate the market power that this institution provides.

Proponents of marketing orders have been known to claim that they promote the general welfare by fostering "orderly marketing." The idea is that consumers would rather see oranges at roughly the same price all winter than to have the price fluctuate from week to week. The achievement of orderly marketing does not of course conflict with the goal of increasing the producers' bargaining power. It is just that instead of bargaining explicitly for price, the administrative committee orders products to be held off the market when a disorderly glut appears. The regulations are printed in the *Federal Register* in a form such as: "The quantity of lemons grown in

California and Arizona that may be handled during the period July 29, 1979, through August 4, 1979, is established at 275,000 cartons." Of course, the producer's disorderly glut is the consumer's bargain, and from time to time, ill feelings are expressed from the consumer side. One possibly useful activity during the Nixon price-control days was the Cost-of-Living Council's efforts to get the committee that controlled the marketing order for lemons to withhold fewer lemons and thereby fight inflation.

Despite the short-run effects of regulating the flow to market, monopoly profits are rendered difficult, probably impossible, to attain because of the lack of control of entry. This is the same problem that was mentioned earlier with reference to cooperatives. The classic cartel model does not apply. So what do the marketing-order powers amount to? Basically, a cartel with free entry, a situation that is analytically analogous to the model of monopolistic competition. Prices to consumers are increased, but the producers' gains are dissipated in cost increases so that the marketing order generates no (expected) profits. Nonetheless, abandoning the cartel would lead to losses for the producers. This is the price we pay for orderly markets.

A more favorable outcome for producers is available if there are alternative uses for the regulated product when it has been held off the fresh market. The alternative use may be a processed product or, for less perishable items like nuts, an export market. Under these circumstances, the regulation of quantities that are permitted on the fresh market can be used to increase the revenue from any given output level by restricting supply for the use in which supply is least elastic, so that revenue-increasing price discrimination involves shifting the output from the fresh market to alternative uses.

Standards regarding grade, size, and quality under marketing orders can be a device for protecting the producer against imports. The Agricultural Marketing Agreement Act requires that imports meet federal marketing-order standards for tomatoes, raisins, prunes, avocados, mangoes, limes, grapefruit, green peppers, Irish potatoes, cucumbers, oranges, onions, walnuts, and eggplant. The Food and Agriculture Act of 1977 added filberts to the list. It was not consumers in search of orderliness who pushed for this addition.

Dairy policy. Milk requires special discussion as a commodity for which not only cooperatives and marketing orders but also price-support programs are important. It is perhaps our most-regulated agricultural commodity (although it is not as effectively cartelized as tobacco or as highly protected as sugar).

First, the price-support program for dairy products puts a floor under the price of milk. This floor, by ruling out contingencies of low prices, provides an inducement to production, and CCC stocks often reach levels that require subsidized disposal. The floor price is sometimes very attractive to foreign producers of dairy products; therefore, imports of foreign butter, nonfat dry milk, and some cheeses are limited by quotas. Even foreign specialty cheeses, which sell for much more than does U.S. cheese at retail, have been subject to import quotas. One of the trade-increasing measures in the multilateral trade negotiations that were concluded in 1979 was a relaxation of U.S. import restrictions on some of these cheeses.

On top of this price-support machinery, a system of federal and state marketing orders has developed, at the instigation of dairy marketing cooperatives. These cooperatives have become controversial because of their recent rapid consolidation into very large enterprises and because of their contributions to politicians out of the producers' substantial political-action funds. The consolidation has caught the eye of the antitrust division of the Department of Justice, which has brought suit against Associated Milk Producers, Incorporated (AMPI); Mid-America Dairymen; and Dairymen, Incorporated. AMPI, the largest milk-marketing cooperative, covers producers in an area stretching from Minnesota and Wisconsin to Texas.

The marketing orders for milk differ from those discussed above in one important respect: they specify a minimum price for milk that is going to fluid use (class 1, which is mainly sold as milk rather than in manufactured form). This price is usually determined with reference to the price for manufacturing milk in the Minnesota-Wisconsin area, which is its most important region of production. Cooperatives may negotiate a price above this minimum for class-1 milk. In each marketing-order area, the quantity that is not sold as class-1 milk goes into manufactured uses. The farmer receives a "blend price," which is separately determined for each marketing order, based on the proportion of milk going into class 1 at the higher negotiated price and into manufacturing uses at a lower market price (which often falls to the CCC-established price-floor level).

The economics of this complex institutional apparatus can be summarized by saying it is a price-discriminating cartel with free entry. Free entry removes the opportunity for rents due to membership in the cartel per se. But the program nonetheless increases returns to producers. The reason is that price discrimination increases the total revenue from any given amount of milk that is sold. The resulting average revenue is a blend price, which is above the price that would clear the market if a single price were charged for all milk of the same quality. Under free entry, producers respond to this higher blend price by increasing their output beyond what their unregulated-market output would be. Farmers increase their output as long as the expected blend price exceeds the marginal cost of producing milk. At the regulated equilibrium, marginal cost equals the blend price. With increasing marginal cost (a rising supply curve), increased economic rents are generated to the farmer-owned scarce resources that are less than perfectly elastic in supply (cows, land, dairy management skills).

Thus, the milk-marketing system has results that are quite different from those of the classic cartel. Too much (from the point of view of the public interest), rather than too little, is produced. Under marketing orders the producers gain, but the losers are not easy to identify. They are: consumers of fluid milk (but not of ice cream, cottage cheese, yogurt, and other manufactured products) and producers of milk who are outside the marketing-order system. These are primarily the producers of low-cost manufacturing milk of the upper Midwest and western New York State. In the ordinary course of competition, one would expect this low-cost milk to undercut the marketing orders because of imports from low-cost areas. However, milk, being mostly water, is heavy in relation to its value. Moreover, it must be refrigerated and kept wholesome for direct human consumption. Thus, shipping it is costly. Even so, the marketing orders explicitly specify an increase in minimum price as the distance from Wisconsin increases, in order to keep interregional competition down. Recent technical advances—most notably, processes for dehydrating milk, shipping it, and then reconstituting it at a distant location—make it possible to supply Wisconsin milk in Georgia, say, at less than local Georgia prices. This challenge has been met to date by restrictions on the use of reconstituted milk. The U.S. Constitution forbids restraints on interstate trade, but ways to limit interregional competition have been found. In the past, some state regulations of milk have gone so far as to require fluid milk that is sold in the state to come from sources that have been inspected by that state's inspectors.

The public-interest issue is to find the losses from the milk-marketing system and allied dairy policies which are not offset by gains to anyone—the net social losses. Some very interesting but difficult problems of concept and measurement arise in this area. The General Accounting Offices (GAO) published a report on marketing orders in 1976, and the Justice Department issued a report on milk-marketing programs in 1977.[2] Both reports were quite critical of the marketing-order system. The Justice Department cited estimates that the net social cost of price supports for milk was about $100 million and of the federal and state systems of marketing orders was in the neighborhood of $200 million. Neither estimate included administrative costs.

The USDA was highly critical of both reports. Basically the same view was taken under the Ford administration (responding to the GAO) and the Carter administration (responding to the Justice Department). In both cases, the USDA made detailed criticisms, along with charges of bias against marketing orders. But the criticisms did not make a convincing argument that the view of marketing orders as essentially being cartelization devices is wrong or that the estimates of net social losses were too large. When the USDA prepared its own report on the impact that marketing-order programs (1975) had had on prices, it found that the programs had had substantial price-increasing effects.[3]

The overall effects of dairy policy on consumers, producers, and taxpayers are impossible to estimate with precision. The price-support program has no major long-term net effects on prices so long as it merely stabilizes by buying and then selling later. The permanent effects result from net additions to demand through domestic and foreign donations of CCC stocks and through import restrictions on dairy products. Donations run at about 4 percent of U.S. production, but the net addition to demand may be substantially less, perhaps on the order of 2 percent (see chapter 2).

The effect of quotas is much more difficult to quantify, because in the absence of quotas, imports would depend heavily on unknown factors, such as the EEC's use of the U.S. market as a dumping ground for dairy products. (For example, the EEC periodically sells part of its butter mountain abroad for 10 or 20 percent of the price that it has paid to European producers for it. If the U.S. would permit it, part of these sales would come here.) Assuming that we would still not allow such dumping, let us say that additional dairy products equal to 2 percent of the total U.S. consumption would be added to U.S. supplies if there were no import quotas.

The assumed 2 percent increase in demand because of CCC donations and the 2 percent decrease in available dairy products due to import quotas have an effect on price that can be determined by the "price flexibility coefficient," which is the reciprocal of the sum of supply and demand elasticities. The relevant demand elasticity is not directly observable. It depends on how the blend-price responds, under marketing orders, to the change in quantities produced, which is a weighted function of demand elasticities for class-1 and manufacturing milk. Econometric estimates of demand elasticities exist for various dairy products at the retail and the farm level. They vary widely but seem to be consistent with an overall value of –0.5. The relevant supply elasticity can be estimated econometrically, but no clear answer has emerged to date. Using 0.3, the resulting coefficient of price flexibility is $1/.8 = 1.25$, which implies that the 4 percent excess U.S. demand that is created by donations and import quotas will generate a 5.5 percent average increase in price. Other recent estimates range from 3 to 7 percent.[4]

Another price effect is due to the two-price schemes of the marketing orders. This effect also is very difficult to estimate, but the available research suggests approximately a 6 percent increase due to higher prices charged on fluid milk (class 1).[5] The resulting economic costs to consumers and taxpayers from a total price increase of 11 percent due to price supports and marketing orders would be about $1.5 billion during an average year. If there were no domestic donations, the costs would all be consumer costs. Consumer subsidies through school-lunch programs, for example, shift some costs to taxpayers. But costs to consumers and taxpayers are not all transferred to producers; some are dissipated in the costs for resources to expand milk production in response to higher prices.

THE PROMOTION AND ADVERTISING OF FARM PRODUCTS

A surprising result of the price-discriminating marketing orders is that they make producers better off in the same way that an increase in demand does, not as a restriction in output does. Farmers as a group are better off, at a given increase in the price of a product, if the increase comes about through an increase in demand rather than a decrease in supply. In the political arena, this fact has led to a search for ways to enable farmers to emulate large nonagricultural businesses by advertising and promoting their products. It is not economically possible for individual producers to do this. Although the problem of free riders is great, voluntary association through cooperatives or trade groups can sometimes be effective.

The American Dairy Association, for example, has for many years promoted the virtues of milk consumption.

The search for ways to finance the promotion of a particular commodity has led to "check-off" programs in which producers or middlemen may vote to assess themselves a small percentage of sales, the proceeds to be used for promotion of the product. Because such programs have proved very difficult to organize on a voluntary basis, trade groups have been lobbying for legislation to authorize the USDA to organize and administer such programs. To protect minority rights, those who are opposed to such programs may petition to have their assessments refunded. Legislation of this type was in effect in 1980 for cotton, eggs, potatoes, wheat, and beef, although two producer referendums have failed for beef.

While one can hardly object if some producers voluntarily join together to promote their product by means of assessments on themselves, it is not clear why the federal government should organize this activity and enforce assessments, or in the case of wheat, why the general taxpayer should help to underwrite efforts to encourage consumers to buy more of a product which the government is at the same time trying to encourage farmers to grow less of. Moreover, purely from the producers' point of view these programs are of dubious value. They are not likely to increase Americans' propensity to eat, but will inevitably have most of their effect in switching consumers from one agricultural product to another. And while one could nonetheless see reasons for publishing information and for promotion to counter the often-misleading claims of those who oppose consumption of certain products on health grounds—for example, beef, milk, and eggs—it is difficult to see this justification for promoting the consumption of wheat, which is perhaps the most highly recommended product by those who object to consumption of so many other foods.

The only clear beneficiaries of the promotion program for wheat are those who specialize in organizing the activities of wheat growers. They will find the demand for their services substantially increased by such a program. Indeed, if one does not push the point too far, the promotion programs encourage the kind of division of interest between leaders or organizers and the rank and file that is often alleged to be characteristic of organized labor. Such division is apparent in the fate of the 1977 and 1980 beef check-off referendums, which failed to obtain the required majorities, despite widespread support of organizations that represent cattlemen.

REPLACING MIDDLEMEN

All these programs have involved increasing the market power of farmers or emulating the behavior of firms in the manufacturing sector in other ways. A more radical recurrent dream among some farm interests is to eliminate the middleman altogether. For example, cooperatives can attempt vertical integration in order to capture the lucrative returns that are supposed to be found beyond the farm gate.

The most interesting recent attempt to legislate the producer past the middleman is the Farmer-to-Consumer Direct Marketing Act of 1976, an effort to subsidize the marketing of food in places other than grocery stores. In fiscal year 1978, the USDA, under this act, paid out $1.5 million in grants to help establish farmers' markets in cities, to get consumers' cooperatives into direct contact with farmers, and for related purposes. Alabama received $110,000 to sample direct marketers and to compile laws and regulations governing direct marketing within the state. Delaware planned to use its $43,500 to help voluntary producer associations, to compile laws and regulations, to develop visual aids, to establish farmers' markets, and to hold a marketing conference. There are fourteen other projects along these general lines.

In 1979, Congress authorized the Bank for Consumer Cooperatives to help these organizations get started. This effort could well turn out to be far more costly to taxpayers than the 1976 act, with similarly negligible benefits to producers and consumers.

The essential fact revealed by these efforts is what costs we are willing to impose on ourselves in order to step on the toes of the nefarious middleman.

REGULATION OF THE FOOD SECTOR

There is a route to smiting the middleman that we have not yet explored. Instead of providing means by which farmers may deal with middlemen from a more powerful standpoint, the state can regulate them directly. Following are a few steps that have been taken in this direction.

Regulation of the price-discovery process. Many farm spokesmen believe that producer interests suffer not so much perhaps from monopsony power in the classic sense as from price manipulation or deceit in the process of contracting a price for products that are sold. Thus, livestock producers brought a legal complaint against meat buyers on the grounds that they manipulate the price reports in the meat trade's "yellow sheet," whose daily quotations are the basis for pricing cattle under many marketing arrangements. Similarly, the big international grain-trading companies are alleged to use their monopolistic access to information and the size of their grain transactions to manipulate central-market prices to their advantage. While proposals have been made to ban or restrict "formula pricing" based on yellow-sheet prices and to have a governmental grain-marketing agency replace some of the function of private grain-trading companies (as has in fact been done in the other important grain-exporting countries), no significant governmental action has been taken to date. And the factual basis for the claims that have been made remains weak. (At the same time [August, 1979] that producers were claiming that yellow-sheet quotations were being manipulated in order to artificially reduce prices, Congressman Rosenthal of New York was charging that the yellow-sheet quotations were being manipulated in order to increase prices.)

The Commodity Futures Trading Commission (CFTC) is an independent regulatory body which is charged with regulating markets in futures, which are closely linked to spot-market prices. In selling (or buying) a futures contract, one agrees to deliver (or take delivery of) a specified quantity of goods of a specified quality at a specified future time and place. The market for buying and selling these contracts is essentially a market for wagers on future spot prices. One reason that such a market is socially useful is that an appropriate wager can reduce the risk for a participant in the spot market. For example, a grain farmer can sell a futures contract for 20,000 bushels at the same time that he commits himself to grow 20,000 bushels. If the market price falls after planting, the farmer's loss on his planted crop is roughly offset by a gain on his futures sale. Thus, the producer is said to be "hedged." It is also socially useful for people who want to speculate to have access to futures markets. The futures market is a great publicizer of information. It rewards the producer of market intelligence while it simultaneously registers for all to see almost immediately the effect that such intelligence will have when it is acted upon.

Some farmers and others fear the manipulation of these markets. Certainly it has been tried. The CFTC is supposed to ensure the proper functioning of futures markets. Unfortunately, there is no clear operational definition of "proper functioning" in these markets. The types of things that are regulated include: maximum permissible size of speculative holdings, limits on price moves in a single day, limits on how much one may borrow from brokers to buy or sell futures, and financial requirements for brokerage firms. It is not well established that these regulations either cause significant harm or do significant good. They do provide relatively stable "rules of the game," which is perhaps the most important consideration for all participants.

The real political threat to futures markets is a tendency to want to shut them down. Trading in onion futures has in fact been banned, and moves have been made to abolish the market in potato futures. The principal beneficiaries of such bans are the large growers and traders who have the best market information. In the absence of futures price quotations, they can keep the price effects of their transactions from so quickly registering publicly that others can adjust their own reservation and bid prices.

Regulation of product markets. The sort of regulation that would respond most directly to market power in the food sector would be antitrust action. However, despite investigations of breakfast-cereal manufacturers and supermarket chains, this area is not currently receiving major policy attention. Ironically, the main attempts to hinder supermarkets—pre–World War II state and local ordinances requiring that substantial special fees be levied on self-service stores and graduated tax rates, increasing with the number of outlets—seem to have been aimed mainly at protecting established merchants from competition from the expanding self-service chains.[6]

The current thrust of regulatory policy is directed not so much at traditional antitrust concerns as at particular problem areas, most of which reflect consumers', not farmers', concerns. These include the accurate labeling of the weight and ingredients of foods, deceptive advertising, clearly understandable pricing, the wholesomeness and safety of foods, and nutritional concerns. The last of these has led to especially bitter controversy, but little as yet in the way of effective governmental action. Such proposals and actions as have occurred are unusual in that they have brought farmer and middleman interests together in opposition to the regulation. While consumer representatives have been the primary proponents of regulation in this area, there is reason for skepticism that consumers in general will be made better off by increased governmental direction with regard to what we eat. Costs imposed on middlemen that do not lead to increased perceived benefits to consumers will make farmers, middlemen, and consumers all worse off.

The most direct regulatory step in food pricing is retail price controls. Price ceilings on food have in fact been imposed, most notably on beef by the Nixon administration in 1973. Whatever their suspicions of middlemen, farmers know that when the maximum that retailers can charge consumers for food is reduced, this will inevitably reduce the maximum that retailers will pay at wholesale, and this in turn will ultimately reduce the maximum paid for farm products. The cattlemen are prone to claim, in fact, that the Nixon price controls caused the dramatic decline in cattle prices that occurred from 1974 to 1976. USDA officials in the Carter administration found this a congenial line of argument. Indeed, in explaining the surge in beef prices in 1979, the deputy secretary of agriculture said: "Today's situation is directly traceable to ill-advised price controls imposed on beef in 1973."[7] While the cyclical behavior of beef prices has more fundamental causes than this, the point is well taken that the well-being of consumers, as well as that of producers and middlemen, may be harmed as well as aided by price controls. It is probably as a result of lessons learned that current inflation-control efforts have resolutely avoided attempting to control food prices directly.

Regulation of input markets. Much of the regulation of input markets returns to the goal of improving the financial position of farmers against input suppliers or nonfarm input users. This is the case with the Interstate Commerce Commission's regulation of rail rates, with allocation rules for natural gas and diesel fuel during the shortages of recent years, and with the determination of interest rates on loans in some states. In loans, however, the development of government programs and quasi-public institutions has been much more important than the regulation of private businesses. In areas where the government itself provides inputs to farmers—notably irrigation water from federal projects and grazing rights on public land—services tend to be provided at rates that are far below market rates. Because these services are provided in limited quantities and are not usually available at the margin, the subsidies result in economic rents for producers and not in lower prices for consumers. (Economic rents are the incomes received by producers above the payments that would be required to induce them to continue to use the same resources in agriculture.)

There are, however, new areas of regulation of farm inputs and production practices which are not directed at producer interests. These include Environmental Protection Agency (EPA) attempts to stem the liquid waste runoff from cattle feedlots, to prevent the soil erosion of fields, to control the runoff of chemicals used in fertilizers, and to keep windblown pesticides from damaging neighboring life forms. The regulations require holding tanks, other new equipment, or changes in production practices, which increase costs. Ultimately, the costs are, like an excise tax, divided between reduced economic rents to producers and losses to consumers through higher prices. The gains are improvements in the quality of streams and the reduction of other external damages. To what extent these regulations can be applied and still have the benefits outweigh the costs is a matter of much dispute on which very little evidence is available.

Regulation of the farm labor market has also increased in recent years, again not with the intention of helping farmers. Farm workers were first brought under federal minimum-wage legislation in 1966, and not until 1978 were the farm and nonfarm minimum wages equalized. Exemptions for smaller-scale employers of farm labor mean that roughly half of the hired farm labor force is still not covered. Increased coverage of farm workers by unemployment compensation and by programs of the Occupational Safety and Health Administration (OSHA) is also a new area of regulation, intended to benefit workers, which adds to farmers' costs. Some of these increases in cost are passed on to consumers, but farmers perceive correctly that they lose income when labor costs increase.

Regulation of the land market is not extensive at present, but many proposals are being considered. Some localities have instituted land-use planning in various forms ranging from toothless statements of desirable patterns of land use to legally enforced zoning. The idea is to reserve areas of good cropland for agricultural use. Some states have instituted programs to aid young farmers in buying farmland by means of subsidized credit.

More emotional attention is being paid to purchases of U.S. farmland by foreigners, which are thought by some to pose a threat of some kind. However, only a tiny fraction of U.S. farmland is owned by foreigners. The USDA estimated that 5.2 million acres of U.S. agricultural land was owned by foreign individuals and groups in 1978, which amounted to about ½ of 1 percent of U.S. land that is in farms.[8] This includes land that is held by U.S. corporations that have more than 5 percent foreign ownership. Nonetheless, some states have placed restrictions on investment in farmland by nonfarm people. And when Merrill Lynch attempted to establish essentially a mutual fund in farmland in which nonfarm investors could participate, the outcry was great enough that the effort was aborted. Who gains by restricting investment in farmland? Those who rent land or intend to buy it in the future. The losers are the present owners of the land. Perhaps one reason that so many farmers favor restricting outside investment is that even those who currently own land intend to acquire more.

4

The Public Interest in Agricultural Policy

An appropriate definition of the success or failure of a policy is essential for judging the performance of the government in agriculture. It is sometimes tempting to appraise policies simply in terms of how well they achieve their goals. Apart from the difficulty of making such assessments, this approach is unsatisfactory, because it assumes that the goals of policy, if indeed they can be identified, are beyond being judged. A law that gives special tax breaks to nonfarm investors in large cattle feedlots, for example, might be accomplishing exactly what was intended by the legislators who enacted it. Yet, in a larger sense, we may not wish to judge the policy a success.

What we need is a concept of the public interest that provides a means for judging outcomes from an independent standpoint. As a first step, one may make calculations, such as the $4.2 billion (table 2.1) budget cost of the major farm programs. But while chapter 2 shows abundantly that U.S. farm programs are costly and complex, this does not mean that they are not in the public interest. One must consider the benefits as well as the costs, and the costs imposed on ordinary citizens as well as costs to the governmental budget, in making a comprehensive measure of the public interest.

THE PUBLIC INTEREST IN ECONOMIC TERMS

The meaning of "public interest" in agricultural policy is best understood by imagining that political decisions concerning the country's food sector are intended to make everyone jointly as well off as possible. Being better off means having increased command over material goods or leisure time or other (immaterial) sources of satisfaction. In the context of the food sector, it means more or better-quality foods, more real income to spend on other goods, or more clean water, open space, or other goods that are jointly produced with food.

The components of well-being derived from agriculture can be identified with, respectively, consumers or producers, or as externalities. The distinction is artificial, in that the same individual may be a producer of wheat, a consumer of foods containing wheat, and a rural resident who is directly influenced by the external effects of food production. Nonetheless, for practical purposes, individuals can be classified according to whether their primary interests are as producers or as consumers.

In this context, making an individual better off means, on the consumer's side, making better-quality food available for a given expenditure, or making a given-quality food available for a lower expenditure. On the producer's side, it means earning greater returns from the production or marketing of food. Making everyone jointly better off means improvements on both the producer's and the consumer's sides. The difficult policy issues arise when one group is made better off at the expense of the other. Unfortunately, this is typically the case in farm policy. How can one judge, in such cases, whether a proposal is in the public interest?

Consider the following criterion: when a policy generates gains to some and losses to others, it is in the public interest only if the gains are greater than the losses. For example, suppose that an embargo on grain exports reduces food costs to U.S. consumers by $100 million but reduces incomes to U.S. producers by $150 million and that it has negligible external effects outside the food sector. Then the embargo would not be in the public interest.

It can be said that this approach is too crude in that it does not allow for differential weights on the two sides. A policy that takes away $2 from Mr. B and provides $1 to Mr. A might be in the public interest if Mr. B is rich and Mr. A is poor. With respect to most matters of agricultural policy, this is not a real issue. There are, of course, many poor consumers of food and many poor farmers; however, there does not seem to be a great deal of difference between the economic well-being of producers and consumers of agricultural products, *as groups*.

In any case, there is a deeper reason for the judgment that taking more than $1 from one group to give $1 to another group is not in the public interest: this is that the losers could, in principle, pay cash to the gainers and make both groups better off without the policy. In the export-embargo case, producers could pay consumers, say, $125 million not to impose the embargo, and both groups would be better off by $25 million than they would have been under the embargo. Similarly, Mr. B could pay $1.50 to Mr. A directly, and both would be better off than they are with the policy. While this approach cannot be guaranteed to be practicable and while it is not always a suitable policy guide, there is usually good reason for being suspicious of a policy that takes away more from one individual or group than it gives to another. Moreover, the difference between the gains of the gainers and the losses of the losers is plausible to use as a measure of the *net social cost* of the policy. Similarly, if a policy results in gains to, say, producers which exceed the losses to consumers and the costs to taxpayers, it will be said that the policy is in the public interest, and the difference is the net social gain from the policy.

The stated purposes of the Food and Agriculture Act of 1977 (see chapter 2)—"to provide price and income protection for farmers and assure consumers of an abundance of food and fiber at reasonable prices"—fall quite naturally into our terminology of public-interest concerns.[1] Benefits to producers are gains in farm income. Benefits to consumers are reductions in food costs. The net social gain is the sum of benefits to producers and consumers, minus the costs to taxpayers. To the extent possible, externalities (such as effects on pollution from fertilizer runoff or pesticide dispersion) should be brought into the accounting, though they involve extremely difficult problems of measurement and evaluation. Having stared them firmly in the face, let us follow Abraham Lincoln's advice and pass on; they are not a major issue in farm-commodity policy anyway. Almost as difficult analytically, but impossible to avoid, are issues of instability and risk. One of the main arguments advanced in favor of governmental intervention in commodity markets is the increased stability in prices and income that they are said to provide. It is therefore necessary to develop a quantitative indicator of the value of stability in order to provide a full and fair assessment of the farm programs.

EFFECTS OF GOVERNMENTAL INTERVENTION IN COMMODITY MARKETS

The most difficult analytical task in judging the consequences of farm programs is to estimate what prices and output would have been in the absence of the programs. In answering such counterfactual questions, one inevitably relies on theory. After forty years of federal farm programs, one simply cannot say what today's agriculture would look like if we had never had them. Simpler, shorter-term analysis of the consequences of, say, increasing a support price by 20 percent also raises difficult analytical issues. The many econometric models of agricultural markets are ill-suited to answering policy questions. Nonetheless, it is possible to make reasonably plausible guestimates of what current interventions are doing to agriculture. To illustrate the possibilities and difficulties, consider the wheat program.

Wheat. After a reduction of carry-over stocks to "pipeline" levels and extremely favorable prices for producers during the period 1973 to 1975, the 1976 and 1977 wheat crops were large enough to generate a substantial decline in price. Measured in 1978 dollars, the real farm price of wheat fell from $6.18 per bushel in the 1973 crop year to $2.47 per bushel in the 1977 crop year ("crop year" being defined as beginning on June 1, the onset of the winter-wheat harvest). These low prices, which caused severe financial problems for wheat growers, constituted one of the principal economic forces behind the growth of the American Agriculture Movement (AAM). The chief result in policy that was attained by the AAM's intensive and disorderly lobbying during early 1978 was an increase in the target price of wheat to $3.40 (from $3.00, where it had been placed only six months earlier in the Food and Agriculture Act of 1977). The loan rate, which puts a floor under the market price, remained at $2.35 per bushel. The average market price received by farmers for 1978 wheat was $2.94 per bushel.

Note that the depressed 1977/78 price was still above the floor price, in real terms. This low support price is the reason for saying that the current program is "market oriented"—CCC loans will not hold market prices above plausible open-market levels in the years immediately ahead. However, the extensive set of production-control measures (see chapter 2) was introduced to increase market price and keep deficiency payments from becoming a political liability through excessive drain on the budget.

The analytical issue is how the U.S. wheat market would have looked in the absence of 1978 programs. There are significant difficulties on both the demand and supply sides. On the supply side, actual production is often significantly different from intended production because of the random element in yields. However, 1978 yields seem to have been about on trend. Even so, there is uncertainty about what production would have been in the absence of the program. It is not correct to assume that the no-program supply curve passes through the 1978 market-clearing point observed in 1978—3.0 billion bushels at $2.94 per bushel. The reason is that the price to which producers were responding was not only the market price, or expectations of the market price, but rather the expected market price *plus* expected deficiency payments per additional unit of production. This suggests that one might assume that the supply curve that generated 1978/79's intended supply should pass through the point at which price equals the target price of $3.40. This assumption would not be an unreasonable approximation except for the complications involving set-asides.

An understanding of the workings of set-asides is important in assessing the producer gains from the wheat program. It is natural to suppose that the payments of approximately $1 billion that were distributed to wheat producers at the end of 1978 represented a measure of their gains from the program. This supposition is false. To see why, consider the revenues of a producer who participated in the program, compared to one who did not. A nonparticipant might have expected to sell his wheat at the season-average price of $2.94 per bushel. A participant could expect to receive $3.40 (more exactly $3.40 multiplied by the allocation factor of .97, or $3.30).

But in order to qualify for deficiency payments, the producer had to participate in the 20 percent set-aside. If the net return from set-aside land is zero, the situation is that the participating producer must have given up the rental value of land that could have produced 0.2 of a bushel per acre. Taking a land rental value of $40 per acre for land that produces an average of 30 bushels per acre, the rental value of land that could produce 0.2 of a bushel would be $0.27. So the price to be compared to the $2.94 is really $3.13, not $3.40. About 60 percent of the apparent gain in deficiency payments is offset by the costs of set-asides.

These costs will vary enormously from farmer to farmer, depending on the availability of inferior land to use in set-asides, the leniency of his county ASCS committee, and the extent of other fixed productive inputs that must also be idled when land is idled. It seems to have been the case that the expected costs were large enough for many producers that it did not pay for them to participate in the 1978 program, sincc about 35 percent of wheat acreage was not enrolled, and the production from this acreage therefore received no payments.

If the preceding calculations are roughly appropriate for the average U.S. wheat producer, the $1 billion of payments really was worth only about $0.4 billion to producers. However, producers also gained (whether they participated in set-asides or not), because the market price would have been less than $2.94 per bushel without the program.

The price effects of the wheat program depend on producers' responses to the expectation of payments, coupled with the acreage restraints required to receive the payments. The difficulties of estimating these effects are formidable. The best available evidence on supply and demand elasticities suggests that the wheat program increased the farm price of wheat by 15 to 20 percent for the 1978 crop.[2] The middle of this range is used in the summary data of table 4.1, implying benefits to wheat growers of about $800 million.

This over-pricing of wheat relative to most other commodities also involves net social costs, in that additional land in wheat production would produce output that would be more valuable than the best alternative uses of the diverted wheat land. This cost, however, was probably much less than the $600 million cost of idling wheat land and not permitting it to be farmed in its best alternative use. Because of the uncertainties in estimating these social costs, the summary data of table 4.1 leave the total social cost at $600 million. For technical details on the concept of social cost, see the Appendix.

Corn. The corn program in 1978 and 1979 involved quite trivial sums in deficiency payments. But as with wheat, there was an effort to idle productive land. From the producers' point of view, the corn program was substantially more attractive than the wheat program, because instead of being penalized if they did not participate in set-asides, they were rewarded if they did. Essentially, it is a matter of the government renting land that is then left idle. The net social cost of the program is the value of products that could have been produced as the implicit rental returns on this land. Since the government paid $160 million to rent corn land in 1978, this is an estimate of the social cost (because farmers wouldn't have rented it if they could have earned more by using it in production). In fact, the government probably received substantially less than it paid for. Studies of past acreage-idling program suggest "slippage" on the order of one-half. That is, if the government rents land that, according to official yield figures, should reduce production by 100 million bushels, actual production will probably be reduced by 50 million bushels.

From the point of view of efficient governmental management of its business, this slippage is not attractive. Strange to say, however, from the point of view of social cost, slippage is good. Indeed, the best result from the social point of view would be for ASCS committees simply to distribute the $160 million to farmers, and then let them do as they please. Given the 50 percent slippage figure, the net social cost of the 1978 corn program would have been about $80 million.

The corresponding cost to taxpayers is $160 million. The cost to consumers is the increase in corn (hence meat) prices caused by the cutback in production, which—given a 50-percent-effective 4 percent set-aside—would be 2 percent. With an elasticity of demand of –.5, corn prices would increase 4 percent, or about 10 cents per bushel, which, at 7 billion bushels, comes to $700 million. The gain to producers would be the $700 million in market-price increase plus the $160 million in acreage-diversion payments, summing to roughly $860 million.

Tobacco. While the tobacco program involves governmental purchases to support the price, it works primarily through control of supply. In 1979, the effective quota that limited the amount of flue-cured tobacco to be marketed was 1.07 billion pounds, which was down 10 percent from 1978. The restriction on supply meant that gains to producers under the tobacco program were financed directly by consumers instead of by the taxpayers. The amount of the transfer can be measured by the rental value of the quotas, which are, with some restrictions, traded openly. The rental value of marketing quotas was about 30 cents per pound in 1978, roughly one-fourth the price of a pound of tobacco. Thus, of the $1.6 billion received by farmers for flue-cured tobacco produced in 1978, perhaps $400 million was a transfer from consumers that was generated by the tobacco program.

It is often said that U.S. tobacco policy is contradictory because it supports the tobacco producer at the same time that the government places restrictions on smoking, produces antismoking advertising, and heavily taxes cigarettes. However, in one sense there is complete harmony. On the consumption side, the policy seeks to discourage the use of tobacco; on the production side, the policy seeks to restrict production. The paradox is in the difference in attitude toward the product and toward the producer.

The difference in attitude is not so apparent in policy toward other issues with regard to health-related consumption. It is not unusual for regulations that have been imposed as a result of ill health among mice to generate substantial losses to U.S. producers who manufacture and use the dyes, preservatives, or whatever it is that the mice ate or were injected with before they became ill. The prevailing view seems to be that such losses should be considered one of the risks of doing business; therefore we are unlikely to see compensation for damages, much less market support or stabilization policies, for the likes of red dye #2 or TRIS-soaked pajamas. Now compare tobacco. It seems relatively certain that tobacco has put an early end to the lives of, not mice, but men, by the thousands. As one might expect, there has been great pressure for a public policy to discourage the consumption of tobacco. On the production side, however, the policy goals are somewhat surprising. The chief aim of tobacco policy is apparently to make it impossible to lose money when one grows tobacco. This impossible goal, ironically, seems to have come closer to being achieved for tobacco than for any other agricultural product.

Sugar. The costs and benefits of the sugar program are obscured by a market that is more volatile even than the grain market. Within the past fifteen years the annual average price in world trade (based on raw sugar at Caribbean ports) has varied between 1.9 cents per pound (in 1966) and 30.0 cents per pound (in 1974), a much wider range than for other farm commodities. U.S. policy has isolated the domestic market so much that the price of sugar at New York on which duty has been paid has been less volatile, ranging from 6.1 cents in 1965 to 29.2 cents in 1974. The argument has been made that U.S. sugar policy has beneficially stabilized supplies and prices for consumers. However, the increased stability has come about by keeping U.S. prices above world prices during periods of low prices. When the supply/demand situation has resulted in high world prices, as in 1974 and 1980, U.S. prices have gone right up with them.

Analysis of the sugar program is also difficult because it is so much in a state of flux. More-recent programs differ from the Sugar Act that expired in 1974 in relying primarily on import tariffs rather than import quotas. In addition, the current program involves CCC loans and deficiency payments, each of which may become more or less important as Congress and the executive branch make further moves in the sugar area.

During most of 1978 and 1979, import duties kept the price of sugar in the United States around 6 cents per pound above the price on the world market. At the restricted equilibrium, the U.S. price is higher by perhaps 5 cents and the world price is lower by 1 cent than they would be without U.S. import restraints. The implied redistributions are: U.S. producers gain $600 million annually (roughly $40,000 per producer), U.S. consumers lose $1,440 million, the U.S. Treasury gains $770 million, foreign producers lose $140 million, which sums to a net social loss, which nobody gains, of $210 million. A final element of the sugar prospect is a proposal to dismantle the domestic program upon the establishment of an International Sugar Agreement (ISA) which would protect the world price and the U.S. price at a level of 15 or 16 cents. This approach would align the United States, the world's most important importer of sugar, against the interests of consumers (importers) and with the interests of exporters! It is as if the U.S. government would urge OPEC to increase the price of oil. The U.S. sugar industry has achieved results that Big Oil could not dream of.

In 1980, world sugar prices increased considerably, as part of general inflation in commodity prices as well as a tighter supply/demand situation for world sugar. The protective effect of U.S. sugar policy, and the consequent redistribution of income, promised to be substantially less in 1980 than the 1978/79 figures given above. The current policy functions much like earlier sugar programs in that it holds the U.S. price up during low-price years but does not provide a corresponding mechanism to hold U.S. prices down during high-price years.

Beef imports. The quantitative restriction of beef imports has been based on average U.S. beef production in the immediately preceding years. A revised formula, applied for the first time to 1980 imports, has been touted by spokesmen for the cattle industry as "countercyclical," in that it permits increased imports when U.S. beef supplies are low and prices are high. These are exactly the circumstances that would draw in more foreign beef under free trade, and it would seem to be even more countercyclical if no quotas were imposed at all. In any case, the Carter administration followed the lead of the Nixon and Ford administrations in suspending quotas in many years and relying mainly on "voluntary restraint agreements" by meat-exporting countries (principally Australia and New Zealand, but also including Mexico and several other Latin American countries). It has been estimated that these agreements keep U.S. beef imports only one or two percent below the level that they would be in the absence of restraints. However, had we been without import restrictions during the 1970s, the market that supplies beef to the United States might well have been better developed today, and we might be importing substantially more than the 6 to 7 percent of our beef supply that is currently imported. Moreover, the United States exports enough cattle, hides, and other livestock products to make us net exporters of animal products. Therefore, a free-trade atmosphere could generate unforeseen benefits to the U.S. livestock industry.

If one supposes that the quantitative restraints on the average hold down U.S. beef supplies by 2 percent and that the tariff adds 2 cents to the U.S. price, the total price effect would be about a

3 to 4 percent increase in the average wholesale price of beef (since the elasticity of demand for beef is probably in the range of –⅔ to –1).[3] The result, taking a 3.5 percent price increase, is a cost to U.S. consumers of about $1.2 billion, of which about 93 percent, or $1.1 billion, goes to U.S. producers, while about $50 million each goes to exporters who gain access to the U.S. market under the restraints and to the U.S. Treasury as tariff revenues. (Incidentally, the allocation of the valuable rights of access to the U.S. beef market has brought the U.S. State Department into the act to help allocate the $50 million among the foreign applicants.)

WHO GAINS WHAT FROM FARM PROGRAMS?

There are numerous estimates of the budgetary costs of the farm programs. Congress requires information on the subject for its appropriations bills, and the executive branch needs estimates for budgetary planning. Any such estimates, however, are conjectural, given our lack of knowledge concerning random exogenous influences on the product markets and concerning how producers will respond to the policy interventions. Moreover, estimates of budgetary costs usually combine such disparate items as deficiency payments, which are a direct transfer to farmers, and CCC loan outlays, which are not (since the CCC acquires grain as an asset to offset the loan outlay). Anyway, budgetary costs are far from being the whole story, as the wheat, tobacco, and sugar programs illustrate.

To get a rough feel for the potential overall economic impact of transfers under current programs, table 4.1 lists estimates for the commodities that are most significantly affected.

In the "ratio of price" column, prices received by producers in 1978/79 are compared to an estimate of what the market price would be in the absence of price supports and production controls. For the major crops, these figures result from the crude calculations outlined above, which provide a rough estimate of income transfers under each program, bringing together some order-of-magnitude judgments. Note that the total costs to taxpayers and consumers exceed the gains to producers by about $1.5 billion. This is a rough estimate of the net social costs of the programs. Given that in 1979 the USDA employed 876 economists[4] and that about an equal number of agricultural economists worked at land-grant (federally funded) state universities, one might hope for more precise and reliable estimates, but they do not exist to my knowledge.

The distribution of gains by commodity depended heavily on

TABLE 4.1

SUMMARY ESTIMATES OF THE EFFECTS OF FARM PROGRAMS IN 1978/79

COMMODITY	PERCENTAGE OF U.S. TOTAL FARM OUTPUT	RATIO OF PRICE WITH PROGRAMS TO PRICE WITHOUT PROGRAMS	COST TO TAXPAYERS[a]	COST TO CONSUMERS	GAIN TO PRODUCERS
				(in millions)	
Wheat	6	1.175	$1,100	$ 400	$1,300[g]
Feed grains	12	1.06	1,020[b]	700	1,500
Rice	1	1.02	10	0	10
Cotton	4	1.02	90	0	90
Tobacco	2	1.33	0	425	400
Sugar	1	1.88	–770[c]	1,440	600
Peanuts	1	1.4	0	200	200
Cattle	21	1.035	–40[c]	1,200	1,100
Milk	12	1.11		1,500[d]	1,200
Wool	0.2	1.3	30	–15[e]	15
All other	40	1.0	0	0	0
Total (or Mean)	100	1.06[f]	$1,440	$5,850	$6,415

[a] Deficiency payments, disaster payments, and diversion payments. Some figures are based on a fiscal year and some on a different commodity-marketing year.

[b] $680 million for corn, $240 million for grain sorghum, $100 million for barley (*Feed Situation*, September, 1979).

[c] Tariff revenues.

[d] This was discussed in chapter 3. Some of these costs are shifted to taxpayers through subsidized milk in school-lunch and other feeding programs.

[e] Gain to consumers through lower prices.

[f] Weighted by value shares.

[g] Payments of $1,100 million plus indirect benefits of $800 minus set-aside costs of $600 million.

the particular events of 1978. For example, rice and cotton received very little benefit in 1978, but both, especially rice, were more affected in 1977. The 1978 estimates suggest that a large fraction of the net income from wheat would have disappeared in the absence of the wheat program. In 1979, however, due to the strength of the grain markets, there were not any significant deficiency payments. The situation for milk changed dramatically in 1980, when congressionally mandated increases in the support price resulted in the accumulation of about $1 billion in stocks of dairy products, implying a price effect higher than that shown in table 4.1.

It is difficult to say which commodities are treated best. Cattle prices are not increased much by tariffs and quantitative import restraints, so that individual producers do not gain much, but the aggregate gains for producers are large. The gains *per producer* are much greater—many thousands of dollars annually—for sugar, tobacco, and peanuts. Dairy producers gain somewhat less. Producers of soybeans, oats, and perhaps barley get virtually nothing, and neither do the producers of the remaining major livestock products—hogs and poultry. The left-out products accounted for 40 percent of the total value of 1977 farm production.

Even the left-out products could benefit indirectly, depending on how various programs were administered. For example, if set-asides come to be relied upon to keep budgetary costs down for wheat or feed-grains programs, then the consequent higher market prices for these crops will indirectly lead to higher prices for soybeans as a substitute crop. By the same token, the higher feed costs would make livestock producers worse off. On the other hand, if production of wheat and corn is increased because of the target-price incentives and if no set-asides are required, the farm program will tend to result in increases in the supplies of these products and thus reduce their market prices. This would tend to reduce the demand for, and hence the prices of, other feed crops. Therefore, if they do not have target-price protection, as soybeans do not have and oats and barley may not have (at the discretion of the secretary of agriculture), producers of these crops will tend to be made worse off. As of 1980, oats did not have a target price, but barley did.

THE DISTRIBUTIONAL CONSEQUENCES FOR OWNERS OF AGRICULTURAL RESOURCES

The gainers from farm programs have been identified as "producers," but in fact not everyone in the farm sector gains equally. One reason is that there are many elements of randomness and arbitrariness in the programs. Consider wheat. The farm price that is used in calculating the deficiency payments is the U.S. average price received by farmers for the five-month period beginning June 1. Since prices typically rise during the marketing year (the year following the beginning of the harvest period), the five-month average typically makes the payments larger than a full-year average would. But this is not always the case. Indeed, the month-to-month variation in the price of grain is so great that the appropriate choice of marketing strategy is often the prime determinant of profit or loss. The same five-month price applies to all producers of wheat, so that the government payment is the same for all producers, regardless of their individual luck or marketing skills. Similarly, farms vary widely in soil quality and other determinants of production costs and therefore produce quite different products. There are winter wheats, spring wheat, white wheat, and durum wheat, some of which have higher quality but a lower yield per acre and thus sell for higher prices on the average. Yet, all producers of wheat receive the same payment on a bushel basis. Thus, there is considerable randomness in regard to who benefits from deficiency payments.

There is a more fundamental reason for some in the producing sector to gain more than others do from farm programs. People receive income as owners of factors of production—inputs into the production processes that generate agricultural output. Owners of different inputs may be expected to fare differently when the price of a product rises. For example, suppose for a moment that increased commodity receipts were passed back to hired workers in the form of higher wage rates. This increase would tend to draw workers into agriculture. This would tend to occur until the net advantages in farm employment became consistent with wages available in other employment. But the pool of labor in other employment is so large relative to employment for a particular farm product that the overall effect on wage rates must be, for practical purposes, negligible.

Similarly, the returns to the labor of the farm operator would probably not be appreciably increased by a program that would increase the price of a particular farm commodity, given time for adjustment. The operators would find, in the first year that the price of the product was increased, that after paying all expenses, they would have more left over to compensate them for their own time. But in the following year, they would divert more effort to this supported crop, and others who previously did not produce the crop would begin to do so, until a tendency would be reestablished for labor returns to be equated in all crops.

The question of who gains from a price-support program thus depends on who owns the resources that will gain most by intensification of effort in producing the supported crop. It seems obvious that the most-favored inputs will be resources whose supply is limited and for which there are no good substitutes. Thus, if expanded production of a crop requires a certain herbicide, the supply of which is fixed, a price-support program for that crop will increase the demand for this herbicide; and because of the fixed supply, this will tend to bid up its price sharply. However, producers of herbicides would then be expected to expand production of the herbicide to the point that their marginal returns would equal their marginal costs. Since industrial production processes of this type tend in the long run to be roughly constant-cost activities, it would be expected that within a few years the price-support program would have increased the herbicide output substantially. While the supply of herbicides is fixed in the short run, it may be almost perfectly elastic in the longer run.

Similar arguments suggest that other chemicals, as well as fertilizer, machinery, and other manufactured goods, will not experience substantial long-term increases in price because of commodity price supports. Thus, the producers of these inputs, like suppliers of hired labor to farms, would tend not to be major long-term beneficiaries of the programs. Nonetheless, they could easily experience substantial short-term losses in profits if the programs were unexpectedly to end or if the programs were to switch in emphasis from payments (output-increasing) to supply control (output-decreasing).

What are the limiting factors in the production of a farm commodity? More precisely, what input is least elastic in supply? Land. When production increases, land that had been better suited to other uses is brought into the industry. Land that is already being used in producing the supported commodity therefore earns rents. If the supported commodity uses only a small proportion of the nation's cropland and if there is a substantial pool of land in other crops that can be moved over to the supported crop, even the gains to land may not be great. The gains to land become dominant when one considers a *system of price supports* for a number of farm commodities. The supply of cropland to agriculture as a whole is quite inelastic compared to the supply of labor and purchased inputs, so that owners of land tend to be the principal gainers from farm programs.

An annual rental charge for land is analogous to the wage rates and input prices that accrue to these inputs. Thus, if a price-support program results in the expected value of one acre of corn production being $220 instead of $200, the preceding argument suggests that the bulk of the $20 increase will show up as increased rental value for the land. If the farm operator owned his land, he would receive a $20 increase in net income. But if he rented the land, then he would have to pay his landlord $20 extra in rent (assuming that the program is expected by all to have the $20 per acre effect). The renter would have to pay more, because the landlord could find other tenants who would be willing to pay $20 more rent for land—that is, competition in the land-rental market puts the gains in the pockets of the landlord. In fact, approximately a third of U.S. farmland is owned by persons who are not farm operators (although many of these are retired farm operators).

The land market that one hears most about is not the rental market but the market for purchase and sale of land. What effects does a price-support program have in this market? When one buys an acre of land, one obtains the rights to a future flow of rental payments. In the case of the corn example, *if* the program is expected to continue indefinitely, the addition to the value of land is the discounted present value of an indefinite stream of $20 annual payments. If the stream were expected to continue at a real value of $20 per year (that is, it rises with inflation) and if the real rate of interest were 5 percent, then the increase in value would be $20/.05 = $400 per acre. Thus, if an acre of land formerly sold for $1,600 (annual rent of $80), it would now sell for $2,000 and rent for $100.

This capitalization of price supports into asset values can be observed most readily in the case of tobacco land, because benefits are tied directly to the quotas that limit the marketings of the crop, which are explicitly tied to land and may be bought and sold with land. Moreover, a quota can be leased (within counties) independently of the land, so that one landowner can rent out his quota to another, and a single operator can enlarge his operation by accumulating quotas that he has leased from other farm operators. Although marketing quotas have no inherent productive value, under the program they function as a fixed input for which no substitute exists. As such, quotas earn rents which provide a market measure of the expected benefits of the program. Market data indicate that a tobacco quota rents for about one-fourth of its sale value. For example, in 1978, a quota that rented for around 30 cents per pound would have a sale value of about $1.20 per pound. The high implicit discount rate on the future rental value of a quota presumably reflects doubt about the future of the tobacco program.

Because the benefits of price-support programs are capitalized into the prices for fixed resources, the real beneficiaries of the programs are quite different from what one might expect at first glance. The gainers depend, not on who produces how much of which commodities, but on who owns the fixed resources that experience increases in price. And it makes a great deal of difference at what time the owner acquired the resource. Someone who bought a tobacco quota at $1.20 per pound, or corn acreage at $2,000 per acre in the earlier example, would capture little of the gains that result from the programs. However, these individuals would suffer capital losses if the programs were ended. Because of the highly leveraged position of many landowners, especially younger farmers (see chapter 1), there exists a great risk of loss if the programs were to end. The situation is analogous to the body's addiction to certain drugs. Our bodies become adapted to an altered state, so that after the initial period of adjustment, the altered state becomes "normal." When this situation has been reached, severe withdrawal symptoms appear if one tries to return to the predrug state. So it is with the position of farmers under commodity programs.

Fortunately, from this point of view, the current programs for the major crops are in such a state of flux and are so complicated and so uncertain as to their effects on the markets that it seems unlikely that they have yet had strong effects on the prices of assets. Even for tobacco, the longest-running and apparently most stable commodity program in existence today, program-created assets that rent for 30 cents sell for $1.20, suggesting a market estimate that the future benefits of the program are in doubt. In this situation, ending the tobacco program immediately would result in a $1.20 per pound unanticipated capital loss for owners of a quota. But prolonging the program for more than four years at present effectiveness would result in an unanticipated capital gain.

In sum, analysis of the input markets in farm-commodity markets requires important readjustment in our estimation of who gains from price supports. It is not the farmers per se, but the owners of the most nearly fixed inputs in farm production. In the short run, these may include many purchased inputs, labor, and specific skills that are involved in the production of a commodity. In the longer term, the gains are expected to devolve primarily on owners of land. So, the issue for distributional purposes is not who produces the products but who owns the land. While commercial farmers are not a class of poor people, as I argued earlier, this statement is even more true for landowners as a class.

The preceding discussion of who the gainers are should make it clear why many farmers would prefer demand-increasing to supply-restricting approaches to support prices. A demand-increasing program increases the use of farm inputs, which generally results in gains to the farmers. But a program that supports prices by restricting supply must reduce the demand for some farm inputs. The gains accrue mainly to specific program-directed factors that are used in order to limit the supply, typically acreage allotments or marketing quotas. Moreover, the set-aside programs do not permit the resources that are released in reducing output to be used in their best alternative uses. This represents an additional cost to producers; it is another reason for producers to prefer demand-increasing rather than supply-reducing programs.

This preference is shared by the nonfarm suppliers of inputs to agriculture. The tractor makers, for example, will never be the rent-earning beneficiaries of supply control. They will only experience a loss in the market for their equipment if production is reduced. On the other hand, the expansion of demand will increase that market. Such considerations may be critical for the processors of farm products, because their operations are often much more tied to a particular area or commodity than are those of the national or international suppliers of fertilizer, machinery, and chemicals. A good example is the processors who convert sugar beets to raw sugar. Suppose that instead of restricting imports of sugar (a U.S. demand-increasing program), we would control U.S. production in order to attain a support price. Because the supply of imported sugar is quite elastic, this approach would require a very large reduction in the production of sugar beets, and certain areas would undoubtedly go out of the business altogether. This would cause substantial adjustment costs for growers of sugar beets, but they could find alternative uses for the land that they took out of sugar beets. However, the processors of sugar beets do not have good alternative uses for their facilities, and they would take very large losses from a reduction in U.S. sugar production. Indeed, their losses from free trade in sugar would probably exceed those of the growers. On the other hand, in the case of sugar cane, the growers do not have such good options for use of the land, so that they would tend to keep producing sugar cane and would bear their losses under free trade, while the cane processors would still have raw material to work with.

In assessing the gainers and losers from price supports, we must look not only vertically along the marketing chain from the farm gate but also horizontally to competing commodities. For example, take the case of a price support for corn: in the Midwest, corn acreage is also typically well suited for soybean production. So a producer will allocate his acreage so that, at the margin, his expected returns from corn and soybeans are equal. When corn production is made more attractive, reallocation of land to corn reduces soybean production and therefore increases the price of soybeans. One might expect, in fact, a tendency for a corn price support to increase the price of corn and soybeans in roughly the same proportion. Otherwise, farmers would keep reallocating land to the crop with the higher return. In this sense it is misleading to describe soybeans as a free-market commodity, because a corn program is indirectly a soybean program. However, the situation would tend to be just the reverse for a production-control program. If the corn price were supported by means of acreage controls, then idled acreage would tend to be switched over to soybean production, so that soybean prices would tend to fall.

In order to forestall program-induced substitution among crops, the current programs have instituted a "normal crop acreage" for each farm. Whatever adjustments a producer makes, he must stay with his normal crop acreage in order to be eligible for program benefits. Set-aside land must be incorporated in normal crop acreage; it cannot be used for raising cash crops such as soybeans. Therefore, there is no possibility for substitution that soybean interests would find harmful under corn set-asides. The determination of normal crop acreage can involve considerable problems. These difficult judgments constitute another of the jobs that have been delegated to the ASCS committees under current programs.

Perhaps the best example of cross-commodity interests in recent farm policy is the relationship between the sugar program and the interests of corn producers. In recent years, technical innovation in corn-based sweeteners, particularly in high-fructose corn syrup (HFCS), has made them a good substitute for sugar. As a consequence of increases in sugar prices during 1974, large investments were made in HFCS. When sugar prices subsequently fell, these investments began to look very dubious. The programs to keep the price of U.S. sugar high have been a lifesaver to the HFCS interests and helpful to corn producers as well. These interests, however, were in the forefront of those who were opposed to achieving the same high U.S. sugar price by means of deficiency payments. The reason is that the protectionist approach would have put both the U.S. producer price and the market price of sugar at 15 to 16 cents per pound in 1978, while deficiency payments to guarantee producers 15 to 16 cents per pound would have generated market prices of perhaps 7 to 8 cents per pound. These prices would have spelled disaster for HFCS interests. Thus, while sugar interests may be relatively indifferent about whether their payments come from consumers or from the U.S. Treasury, the corn interests most assuredly are not.

On the losers' side of the ledger, in addition to costs to taxpayers, there are costs to consumers through high food prices. A rough idea of the effects on price at the farm level is given by the figures in table 4.1. It should be emphasized, however, that these estimates are very uncertain. Also, producer-price increases that are accounted for by deficiency payments do not increase the market prices; indeed, they tend to reduce them. My "best guess" is that the farm-commodity programs of 1979/80, together with other regulatory programs that were discussed in the preceding chapter, can be expected to add about 4 percent to the overall level of farm prices and about 2 percent to the overall consumer retail cost for farm-based products. This would amount to about a $5 billion annual loss to consumers.

An alternative estimate can be inferred from recent statements by officials of the USDA. In 1977, before the 1977 act was in effect, U.S. net farm income was $20 billion, while in 1978, when programs under this act were in effect, net farm income rose to $28 billion. In their speeches to farm groups around the country, the secretary of agriculture and other officials of the USDA attributed substantial gains in farmer well-being to programs established by the 1977 act, especially to the role that extended loan and set-aside programs played in raising grain prices. However, they have given no quantitative estimates of the effect. If half of the gain in farm income were attributed to the programs, this would amount to $4 billion.

When it comes to the consumer side of the ledger, USDA officials have not been so emphatic about claiming credit for the increase in food costs during the same period. In mid 1979, the USDA issued a paper entitled "Food Prices in Perspective" to explain why food prices had risen 10 percent in 1978 and an average of 9 percent annually since 1973, but it did not find farm policy worth mentioning as a causal factor. In fact, the $1 billion increase in farm income that is due to deficiency payments is not passed on to consumers. But for gains in farm income that are reflected in higher market prices, we may be confident that much of the farmers' gain is the consumers' loss. The farmers' gains must come either from middlemen (being absorbed in lower returns to marketing services) or from the ultimate consumers. Because the supply of marketing services is quite elastic in the long run, returns in marketing probably are not greatly affected in the long run, for the same reasons that returns to manufacturers of purchased farm inputs are not greatly affected by farm programs. Therefore, it can be expected that after an initial adjustment period, the farm programs essentially involve transfers from final consumers to landowners, with food retailers, processors, distributors, assemblers, and farm operators themselves (as owners of labor and management) serving primarily as conduits through which higher returns are passed back to those who have acquired ownership rights to the productive properties of the soil and other specialized agricultural resources.

THE SOCIAL COSTS OF CURRENT PROGRAMS

The identification of gainers and losers from farm programs is very difficult, but it is necessary if one is to understand why we have these programs and what their effects are. Still more important, and more difficult to estimate, are the social costs mentioned earlier—the costs that consumers and taxpayers give up but that *nobody* gains. The earlier discussion suggested social costs that added up to perhaps one and a half billion dollars in 1978 due to restricted U.S. consumption. Yet these calculations leave out much that constitutes the real social costs of our farm policies.

The social costs result, in the cases considered, because programs cause too high prices and too little consumption. But in some cases where the programs have a small or not easily determined effect, there are still real problems.

Consider wheat. Many programs impinge on the wheat market, with a bewildering array of incentives and counterincentives. Target prices encourage increased production, while set-asides simultaneously try to hold production down. At the same time that we are trying to control production, we have government-sponsored research to increase yields; a research and promotion program to encourage domestic wheat consumption; a federal grain-inspection program, as well as other measures to encourage exports; and the CCC loan program and storage-construction loans to increase stocks. The target-price program encourages increased yields per acre by means of the increased use of fertilizer, pesticides, and other inputs; whereas the disaster-payments program encourages farmers to exercise less care in production practices generally, and so encourages lower yields. Finally, we have P.L. 480 and CCC export credit to boost demand further, and a menu of voluntary diversion and other production-control authorities which are available at the discretion of the secretary of agriculture, together with his authority to cut loan rates, manipulate storage subsidies and release prices, purchase surplus commodities for school-lunch programs, impose export subsidies, make payments for wheat acreage that is grazed out, provide emergency relief, and establish import controls. Such is the program that has been advertised under the label "market-oriented."

The simultaneous attempts to reduce supply and to increase demand could possibly result in the same output of wheat as if there were no programs at all. But we cannot therefore assume that these programs may be simple transfers of income with no resource misallocation or social costs. The net social costs of our farm programs include the funds and manpower expended in concocting and administering these programs as well as the time and effort spent by farmers in attempting to become informed about them, to comply with them, or to subvert their intention, and the costs incurred when political representatives of farm interests engage in congressional lobbying for and other promotion of their legislative favorites.

While it is not possible to provide a precise estimate of these costs, they are quite likely to equal the conventional measured social costs. For example, counting ASCS employees, members of county committees, related USDA and congressional staff, and commodity-group analysts, probably at least twenty thousand work-years are devoted to farm-program affairs. At the ASCS average compensation of $15,000, this would amount to $300 million annually, which we probably ought to increase by 50 percent for overhead costs.

With administrative costs added to the figures in table 4.1, the estimate of net social loss due to the farm programs would be increased to the neighborhood of $2 billion. The following line of argument suggests a much higher estimate.[5] Producer interests have been able to persuade the U.S. government to transfer roughly $6 billion to them from their fellow citizens. Given rewards of this magnitude and given the fact that lobbying and other political activity is open to all, one would expect roughly comparable amounts to have been spent in obtaining the transfers. Moreover, since it would be worth $6 billion to nonagricultural interests to have these transfers end, one might expect a tendency, on the part of nonagricultural interests, toward willingness to spend comparable sums on political activity. While this is a conjectural and empirically unconfirmed calculation, this reasoning suggests that we ought at least to consider the idea of counting the whole $6 billion that is transferred as net social loss.

On the other hand, some agricultural economists have argued that farm programs entail social benefits that have not yet been mentioned. These arguments are not entirely the analytical joke that they are sometimes taken to be. They demand serious attention.

5

Objections to the Criticism of Farm Policy

In terms of benefit-cost analysis, the existence of net social costs implies that the social rate of return to investment in farm programs is not only small, it is negative. Going even further, the approach implies that any programs of the type that are currently in operation —programs that attempt to restrict production, put quotas or tariffs on imports, or subsidize producers by means of deficiency payments —entail net social costs.[1] The only empirical issue is how large these costs are.

Although such measurements of net social cost are common in economists' treatments of farm programs and other market interventions, fundamental objections to them have been raised. Let us begin with the following assessment by an eminent agricultural economist, George E. Brandow:

> Economic evaluation of social costs associated with particular policies has frequently employed partial equilibrium analysis and the concepts of consumers' and producers' surplus. . . . Reservations on theoretical grounds often are based on a reluctance to agregate personal utilities and on second-best considerations.
>
> This reviewer is unwilling to aggregate personal utilities indiscriminately. He is particularly unwilling to accept the assumption that there exist empirical counterparts of either the perfect competition situation or the equivalent situation under the constraints of a program. As the general model of the agricultural sector discussed earlier indicates, agriculture has been and is in chronic disequilibrium. . . . The neat alignment of resources, output, and prices specified by the perfect competition model is far from duplicated in free markets, and the equally neat alignment assumed under the constraints of a program is not experienced when programs are in effect.[2]

My understanding of Brandow's essential points is that: (1) It is inappropriate to "aggregate personal utilities indiscriminately," and

(2) programs may be "second best," because in the absence of them, we would observe, not a socially optimal competitive equilibrium, but "chronic disequilibrium."

To "aggregate personal utilities indiscriminately" presumably means to take $1 for farmers as being equivalent to $1 for consumers and $1 for taxpayers. However, as discussed at the beginning of chapter 4, when a net social costs exists, it is, in principle, possible to make *all* parties better off, so that beneficial reform is possible no matter how utilities are aggregated. As a USDA official is reported to have said about an anticipated price boost under the sugar program: "For what this program costs consumers, you could buy all the sugar growers out."[3] A roughly similar but more precisely defined statement could be made in any instance where a significant net social cost exists. Yet it may be nearly impossible in practice to achieve the appropriate compensating mechanism to make everyone better off by ending a farm program.

If we open the door to attempts to justify farm programs by discriminating among personal utilities, the problem immediately arises as to how the necessary discriminations are to be made. The justification for transfers of wealth from taxpayers to farmers involves some implicit aggregation, or some ability to put quantitatively on the same scale the utilities of the gainers and the losers. The only alternative to doing it indiscriminately is to discriminate. On what basis shall we become biased in favor of farmers?

Brandow's second point fits in with the idea, which was widely held in the 1950s and 1960s, of a "farm problem," which it should be the goal of agricultural policy to solve. The problem is chronically low incomes in agriculture. The reasons that have been suggested to explain this situation include the monopsony power of middlemen, the tendency for a smaller share of income to be spent on food as an economy grows, more rapid technological advance in agriculture than in nonagricultural production, overoptimistic investment in fixed capital by farmers, inflexibility in adjustment of the farm labor supply to declining demand for it, and lack of schooling and skills among farmers and their families. The dominant idea is that chronic disequilibrium existed in the sense of having too many resources in agricultural production.

"Disequilibrium" is a tricky term. Not only yearly prices but even weekly or daily prices are in a sense equilibrium prices in that they are all statistical artifacts of a market-clearing process. They are all estimates of points in price/quantity space at which supply equals demand. The essence of the disequilibrium idea is that today's market-clearing price is different from some normal long-run equilibrium price that is justified by underlying conditions of supply and demand. In such a situation, the short-term equilibrium position may indeed involve a departure from what is socially optimal, and so it may not be the proper point of comparison for calculating the social costs of programs. Nonetheless, price-support programs do still have quantifiable net social costs, or possibly social gains, although measurement of them requires knowledge of the long-run supply function (see Appendix).

With information about the long-run supply curve, standard welfare economics can quantify the net social costs of the failure to adjust to equilibrium. In the short-run disequilibrium situation, the long-run opportunity cost of resources used in farming is greater than their returns in agriculture. By adjusting output to a lower level, a net social gain can be obtained. However, it may be costly to make the adjustment: indeed, it must be if producers persist in using their resources to yield returns that are below their long-run opportunity return.

How might the farm sector arrive at, and remain chronically at, a position of disequilibrium in this sense? The combination of rapid technical progress in agriculture, coupled with general economic growth and a small income elasticity in demand for food, can indeed require continuous shifts of resources out of the farm sector. But this has not occurred so suddenly as to engender a large unanticipated shock to the system.

Considerations such as these long ago led observers of the farm problem, such as T. W. Schultz and D. Gale Johnson of the University of Chicago, to concentrate on factor markets as the key to solving the farm problem. Even if rapid technical progress in agriculture, coupled with a low income elasticity of demand for food, decreases the demand for farm inputs, a chronic disequilibrium requires some hindrance to adjustment in factor supplies. An example is the immobility of farm labor. Thus, discussion of the farm problem came to focus on the skills, off-the-farm opportunities for work, and the ability of the agricultural labor force to adjust to disequilibrium. The idea of low returns being due to an excess supply of labor is borne out in the rapid and continuous decline in the farm population, which still continues. Moreover, the adjustments to equilibrium have indeed been in the factor markets, not in the product markets. Agricultural inputs *as a whole* have not declined, and agricultural production has increased. Capital inputs in agriculture have increased at the same time that the labor supply has been decreasing. Land has gone out of production in some areas at the same time that it has come into production in others, the latter instances being mainly land that has been created by capital investment—for example, because of irrigation or drainage projects. These considerations confirm that the farm problem was always to be found in the input factor, particularly the labor markets. Remedies that are aimed at product markets were able to provide, at best, temporary relief, and over the longer term, they actually hindered necessary adjustments in the labor market, to the extent that they had any impact at all. Production-control programs in this light look a little better; but by concentrating on reducing the land input (which there was no long-term reason to reduce), they also provided an artificial stimulus to labor-intensive production, the wave of the past.

Finally, however serious the farm problem was during the 1950s and 1960s, it does not seem to exist today, at least in the form that it did then. The massive adjustments of the 1950–70 period were associated with gains to the owners of farm inputs, but quality- and skill-adjusted total rates of return to labor and capital do not seem today to be unduly out of line in the farm sector compared to the nofarm sector. Reduction in the use of labor in agriculture continues to occur. But this phenomenon seems to be as much attributable to the pull of the invisible hand of off-the-farm opportunities as to the push of the invisible foot of subpar farm returns. A nice illustration of the extent of the adjustments that have been made is available in the work of Kaldor and Saupe,[4] which Brandow cites as one of the best demonstrations of the earlier chronic disequilibrium in agriculture. While any estimates of this sort are open to errors from many sources, Kaldor and Saupe use 1959 data to suggest that an "income-efficient" commercial agriculture in the North Central States, projected to 1980, would involve about one-third of the labor that then existed, with farms being about four times as large in terms of acreage, while real prices of products were projected to fall with continued technological progress. These changes were seen as a solution to the farmers' low-income problem.

As we entered the 1980s, changes of this order of magnitude had in fact occurred, and the chronic problem of disequilibrium appeared in fact to have been solved. There is irony in the fact that Secretary of Agriculture Bergland then saw the solution to the old farm problem as itself a problem, a "structural" problem. It would be a fine irony indeed if, having done virtually nothing to cure the old farm problem, the federal government should act to negate the slow, painful solution that the labor markets finally provided.

Nonetheless, even if the farm problem as conceived in the 1950s and 1960s is not relevant to current policy and even if social-cost accounting is not so dependent on restrictive assumptions as is sometimes supposed, there is still the matter of Brandow's "second best" argument, his contention that the absence of farm programs would result in a situation that would entail social costs even larger than those caused by the programs.

The standard of comparison in social cost analysis is supply-demand equilibrium in an unregulated competitive market. The question is whether that is an appropriate standard. What are the underlying forces in U.S. agriculture that would come to the fore under alternative policy regimes, particularly laissez faire? Economic and political arguments have been put forth that policies that are perhaps not identical with but certainly not radically different from current approaches are socially preferable to the available alternatives. And it may be argued that the appropriate remedy to deficiencies of current policies is not less market intervention but better, more intelligent, and perhaps more comprehensive intervention.

What is the point of view, the theory of the economics of agriculture, that makes it a plausibly productive activity of the U.S. government to regulate agricultural production, marketing, and prices? Among the many arguments that have been advanced, the following seem to have the most appeal: (1) the theory that unregulated markets for farm products cannot cope adequately with instability; (2) the theory that agriculture has other special characteristics that unregulated markets cannot properly handle; (3) the theory that farmers cannot obtain fair returns in the unregulated marketplace; (4) the theory that governmental intervention is necessary to countervail the actions taken by other governments; and (5) the theory of the threat posed by large-scale, mechanized farming.

THE THEORY THAT UNREGULATED MARKETS CANNOT COPE ADEQUATELY WITH INSTABILITY

Proponents of commodity programs often put considerable weight on the notion that consumers would prefer to be faced with stable rather than with fluctuating prices. It is certainly true that consumers would prefer stable to rising prices, but this should be sharply distinguished from random fluctuations around a fixed mean. An economist with the USDA, Frederick Waugh, showed in 1944 that there were good reasons to expect consumers actually to be able to do better under fluctuating rather than under stable prices.[5] The essential reason is that consumers generally will not simply buy the same quantities at high and low prices but will adjust their purchases to buy more than the usual amount of a commodity when it is unusually cheap, and less than usual when it is dear. In this process, consumers can, and normally will, obtain more total benefits per dollar spent on food when prices vary than when prices do not vary.

Waugh's argument is as relevant today as it was thirty-five years ago, perhaps even more so in that a greater variety of more storable foods is available today, so that consumers may more readily adjust purchases to short-term changes in prices at the supermarket. In any case, the appeal to consumers' interests in stability as a justification for farm programs is highly dubious.

It is still possible to maintain, however, that there are opportunities for consumers to gain from stabilization because uncertainty per se is undesirable. For example, poor people and consumers who are strongly averse to taking risks may suffer enough losses of wealth during periods of high prices that they would prefer stability. It is also widely accepted that extreme price shocks that are associated with extraordinary worldwide production shortfalls, such as occurred from 1972 to 1974, cause macroeconomic dislocations that are best viewed as external costs of instability in grain prices. Since a private, profit-seeking grain-storage industry could not capture the benefits of these external shocks (especially when governments act to prevent "windfall profits" from accruing to the stockpilers), there is a case for a public storage/stabilization program.

At the low end of the price scale, arguments of externality have also been put forward. The extreme form of this position is that if prices are allowed to fall in times of large supply, farmers will go out of business, after which large corporations will take over agricultural production, and then prices will go through the roof. A more moderate version simply recognizes that because of inelastic short-term demand and supply, coupled with inherently unstable farm production due to weather, extreme fluctuations in prices will tend to occur that will impose severe hardship on marginal producers and great short-run stress on all producers. To the argument that private speculators in commodity storage can operate profitably to reduce such instability to the extent that it is justified by the costs of storage, one can answer that the resulting price variability has in the past still been too large and that proper policy can prevent needless hardships that are associated with price extremes.

If producers are averse to taking risks, there are external benefits to be had from governmental price stabilization, even beyond the point at which expected price gains cover the costs of storage (i.e., storage beyond what a profit-seeking private storage activity would generate). Further storage during periods of low prices could be expected to lose money, after storage costs (including interest) have been substracted, when grain is sold during periods of high prices, but the resulting price stabilization might induce risk-averse producers to increase output (at given mean product and factor prices) and might therefore result in a combination of increased economic rents to producers and lower prices to consumers that would justify the costs of storage. However, whether the supply response to reduced instability in prices is enough to cause social benefits that exceed the costs is open to question. Some economists have claimed that historical experiences, such as that of the potato program of the 1940s, demonstrated the dramatic effects of reduction of risks:

> The price support program created an unlimited market for potatoes at approximately the average price (in relation to parity) which had prevailed for 30 years and virtually eliminated the risk of very low prices. . . .
>
> The response to the program was immediate and of such magnitude as to be unmistakable. . . . *The expansion occurred through production adjustments induced by—motivated by—the greater price certainty.*[6]

This argument is a logical howler. Almost certainly, the increase in mean price resulting from the elimination of the lower half of the price distribution had a lot more to with the expansion of output than did price stability per se. If it were strictly a matter of stability, roughly the same results could have been achieved with a price *ceiling* at the thirty-year mean. Recent studies that have held the mean price constant in econometric models have also found that there would have been an increase in output when prices were less variable, but the quantitative effects would have been much smaller. Moreover, even these results are questionable, because even with a given mean price, reduced variability can lead to increased mean revenue (and costs to consumers): for example, when the demand curve is linear. In short, there is no solid evidence that pure stabilization via storage leads to net social gains because of risk aversion among producers. Still, there *may* be such gains, and it might therefore be worth undertaking price- or income-stabilizing policies in order to achieve them. These possibilities will be explored further below.

THE THEORY THAT AGRICULTURE HAS OTHER SPECIAL CHARACTERISTICS THAT UNREGULATED MARKETS CANNOT PROPERLY HANDLE

Food undoubtedly has special characteristics as an item of consumption, and farming has special characteristics as a productive activity. The policy significance of these characteristics, however, is not always apparent. Indeed, the economists who study the economy as a whole tend to see agriculture as an industry that is essentially like any other—there are differences, but these are superficial, while the fundamentals of economic behavior are the same. People who make a living in agriculture tend to see its unique aspects as fundamental, generating problems that require special analytical approaches and remedies. Economists who concentrate on agriculture see elements of truth in both positions.

One of the most notable of the special characteristics of agriculture is that farm production is subject to long lags and uncontrollable fluctuations because of the biological nature of its production process. Another is that the extreme perishability of some products creates special difficulties in handling and marketing them, so that these products often go to waste. However, it is possible to overstate the uniqueness of agriculture on these points and to underestimate the ability of the market mechanism to send the appropriate production and marketing signals. No agricultural producer's income depends on a product that is more perishable than, for example, the labor services provided by farm workers in harvesting vegetables. They offer services of labor time which, if not utilized immediately, depreciate instantaneously to zero value. Such perishability is inherent in the provision of some agricultural products and some labor services, and this results often in products and services that perish. Provision of market information may prove valuable in these situations, but how commodity programs can provide better results is not clear. Indeed, for many of the most perishable commodities, no programs exist, and there is no apparently strong demand for them.

In the context of the marketing of dairy products, Secretary of Agriculture Butz once expressed the view that although he supported a market-oriented agriculture, an unregulated market was not capable of appropriately allocating to unprocessed and processed (more storable) uses a geographically dispersed product (such as milk) the supply of which was unstable. Although the problems of coordination are great, there is no well-developed theory of how, from the viewpoint of public interest, a regulated market can be expected to handle these difficulties better. The uniqueness of milk is not clearly relevant to policy, nor does it provide an a priori case for the existing regulatory programs.

There is, in fact, an essentially unregulated market that has the key features of milk—a highly perishable raw product that has a variable supply, a geographically dispersed production, and a mixture of relatively unprocessed and highly processed uses—namely, eggs. While it is not claimed that the unregulated egg market is a model of perfection, there is no evidence that the milk market has performed enough better to have justified the higher consumer prices and hundreds of millions of taxpayers' costs involved in U.S. dairy policy. Indeed, despite assertions of the uniquenesses and vulnerabilities of the dairy industry that simple cost-benefit analysis cannot capture, there is no better bottom-line assessment than the type of calculations of social cost that were reported in chapters 3 and 4 with reference to dairy programs.

More should be said, however, about time lags in adjustment to changing circumstances that have been caused by biological growth patterns and the necessity to invest in long-term equipment in an uncertain market environment. This aspect of instability is distinguishable from both the "chronic disequilibrium" and the short-term variability that were discussed earlier in this chapter. I believe that Vernon Ruttan had this aspect of agriculture in mind when he said: "The rationale for public intervention in agricultural commodity markets is, and will continue to be, essentially similar to the rationale for setting rates and regulating output in the public utility and transportation industries, that is, to lend stability to an industry which technological and economic forces would render chronically unstable in the absence of such intervention."[7] The problematical element is the risk of loss on irreversible fixed investments. There is no doubt that market signals can lead to large private and social costs under these circumstances. (Whether governmental intervention is a promising remedy is another issue.)

Perhaps the most important point to make about many of agriculture's special characteristics is that, over time, they are becoming less special.[8] Managerial skills and risk-management techniques that have been developed in financial and industrial contexts are being increasingly applied to farming. Mobility of resources has increased enough that nonfarm opportunities for use of the farmer's labor and capital can cushion the shocks due to economic change that must be absorbed. This new situation greatly diminishes the policy relevance of agriculture's special characteristics.

One special feature of agriculture persists, however. It is the nearly atomistic competitive nature of agriculture in contrast to the concentration of firms in the service and processing industries that buy from and sell to farmers.

THE THEORY THAT FARMERS CANNOT OBTAIN FAIR RETURNS IN AN UNREGULATED MARKETPLACE

Perhaps the chief reason for nonfarm people often being sympathetic to farmers' pleas for governmental aid is a belief that farmers tend to earn unfairly low returns for their efforts. As the data in chapter 1 indicate, however, income levels and rates of return to investment have actually tended to be higher in commercial agriculture than in the nonfarm sector.

Nonetheless, there is evidence that both food retailers in concentrated markets and food manufacturing concerns are able to charge prices and to earn returns that exceed purely competitive levels.[9] The demand for agricultural products is derived from the marketing sector's demands for these products to be used in food- and fiber-manufacturing and in sales to final consumers. The relevant demand curves are those derived from the profit-seeking behavior of individual enterprises in the food sector. The fact that final-product output is determined by equating the middlemen's marginal revenue to their marginal cost (made up of the cost of raw farm products and of nonfarm inputs that are used in marketing) results in a derived demand for farm products that is less than would occur under competitive pricing (see the Appendix). The result is a transfer from farmers and consumers to middlemen, and this involves a net social cost—a loss to farmers and consumers, which nobody gains.

When economists at the University of Wisconsin (cited in note 9 of this chapter) published their estimates of the effects of monopoly power, the estimates were criticized as being much overstated. The estimated price-enhancing effects amounted to sums greater than any monopoly profits discernible from the income statements and balance sheets of food corporations. And if it is so great to be in the food business, what happened to A & P, Food Fair, and other corporate casualties? The most plausible hypothesis is that rents to monopoly power are bid away by competition in the capital markets.

The prices for the assets and for investment opportunities of firms are bid up to the point that the rate of return is comparable to the economy's general rate of return to capital. Random deviations in returns should then lead to roughly the same frequency of bankruptcies in the food area as in other businesses. Nonetheless, prices remain at the imperfectly competitive levels. The situation is very much like that of the "cartel with free entry" interpretation of cooperatives under marketing orders (which was discussed in chapter 3).

Perhaps as important as monopoly power is monopsony power—the ability of a single buyer or an organized group of buyers to influence the farm price for a commodity. In this area we do not have empirical studies such as the ones that the University of Wisconsin group has undertaken for food manufacturing and retailing. But since marketing services that are available to cattle-producing areas are even more concentrated, with fewer substitutes being available, than is the case in food manufacturing and retailing, monopsony power is very likely to exist. It has essentially the same effects as food monopoly power has on farmers and consumers—less is purchased from farmers than would be purchased under competition, resulting in lower farm prices and higher consumer prices, the difference going to monopsony profits. Again, the monopsony profits may be dissipated by investment rivalry, but net social losses may nonetheless persist.

How does all this fit into the social-cost analysis of farm policy? It supports a second-best argument in favor of governmental intervention. A policy that would guarantee an appropriate price to farmers by means of deficiency payments without production controls could result in increases in output sufficient to offset the restricted demand of monopolistic middlemen. However, a production-control approach under imperfect competition does not have any desirable second-best properties. It raises the price that is received by farmers, but it makes consumers even worse off than they were under the unconstrained monopoly situation. Thus, from the point of view of the public interest, production-control programs are even worse under a monopoly food sector than under competition (for elaboration of these points, see the Appendix).

The same is true for policies that attempt to remedy a food-sector monopoly by granting monopoly power to associations of producers. If the associations can control production, they can capture returns that would otherwise accrue as monopoly profits in the food sector. But the main loser is the consumer, who pays still higher prices. In short, if we arm producers with better weapons in their struggle against imperfectly competitive middlemen, the expected result is that they will jointly exploit the still-unarmed consumers. I conclude that traditional farm policy, with the possible exception of deficiency payments without production controls, is not plausibly rehabilitated on second-best grounds because of imperfectly competitive middlemen. The proper policy approach is to deal with monopoly power by means of antitrust remedies that are congruent with those used in other sectors of the economy.

THE THEORY THAT GOVERNMENTAL INTERVENTION IS NECESSARY TO COUNTERVAIL THE ACTIONS TAKEN BY OTHER GOVERNMENTS

The most important countries in importing and exporting farm products impose more centralized direction of trade flows and price determination than does the United States. It is said that in such a world, the U.S. is at a disadvantage because of its reliance on a decentralized market-pricing system. The truth of this statement is not so obvious as may be supposed. For example, if state-directed marketing boards in grain-exporting countries hold back supplies in order to raise prices, the unorganized grain-exporting producers become free riders on the cartel. There is a real possibility, however, that U.S. producers and consumers might be able to benefit jointly through monopolistic exploitation of the world markets for grain and soybeans, since the United States is not strictly a price taker in these markets.

Distinct but related issues arise from the use of subsidized exports as a way of disposing of surpluses that have been created by price-support programs abroad. The countries of the EEC are notorious for using such practices. Japan is exporting rice, even though the domestic farm price there is about three times the world price. Moreover, it is true that some production costs abroad are lower than costs in the U.S. because producers in other countries may not have expenses caused by regulations such as bans on pesticides.

In addition, highly protectionistic policies in importing countries may reduce the market for U.S. products. These policies raise problems for the analysis of social costs that compare the situation with free markets. There may not be any free markets, even in the absence of U.S. programs. The approach followed in this book, as in the discussion of dairy imports, is to compare U.S. programs with a situation in which our domestic programs are changed while foreign economic policy and our responses to it (e.g., countervailing duties) remain in place.

In these areas, there is an obvious opportunity for useful government-to-government confrontation and negotiation. Discussions of trade policy, such as the General Agreement on Tariffs and Trade, represent a response by governments to problems created by governments. The market is quite helpless in these matters.

THE THEORY OF THE THREAT POSED BY LARGE-SCALE MECHANIZED FARMING

The secretary of agriculture recently said: "Today, the 200,000 largest farms account for nearly two-thirds of all agricultural production. In contrast, as recently as 1960 small farms with less than $20,000 in sales produced nearly half the value of all farm products. Today, farms of that size produce less than 11 percent of our farm output."[10] It must be noted, first of all, that this comparison is misleading, because a $20,000 farm today is less than half the size that a $20,000 farm was in 1960, since the dollar has depreciated by more than half. No government official should be permitted to slide over that fact. Nonetheless, the secretary's general point is valid. More interesting, however, is his attitude about the situation:

> I am deeply concerned about what I see happening to the structure of agriculture. I am deeply concerned also about why it is happening. And I am concerned most of all with the desperate need to ask ourselves if what is happening is what we want . . . or what the Nation truly needs. The truth is, we really don't now have a workable policy on the structure of agriculture. To the extent we talk about such a policy, its focus is always on the *number* of farms. But on what basis do we decide whether we should have 1 or 3 million farms? Surely, it is time to develop a national farm structures policy.[11]

Surely it is time to consider whether governmental activity in this area is wise. Can it be expected to have benefits that are greater than the costs? Beyond this, the secretary's reflexive viewpoint that all problems are policy problems is a cause for fundamental unease. On what basis is the occupation of the two million people who make up the difference between the figure of one and three million farms something that "we" should decide through governmental action?

The point here is the theory that large-scale agriculture poses a threat. The view of U.S. agriculture of which this theory is a part seems not to be as widely held as the others that have been discussed, but it does have strong adherents besides Secretary Bergland. Consider the following position, which goes well beyond the secretary's: "Contrary to the popular assumption that food was among our last competitive industries, food growing, processing, and marketing have fallen increasingly under the monopolistic control of a few banks, conglomerates and holding corporations. Aided by the Federal Government, these corporations have been permitted to regulate food supplies for domestic consumption as well as encouraged to divert substantial quantities of crops for foreign trade."[12]

This oligopolistic theory of agricultural production is not congruent with the available data. The 1974 Census of Agriculture surveyed corporate-owned farms and found that there were 28,442 of them, a substantial number. However, most of these were family or other small, closely held corporations, not the giants that the oligopolistic theory is concerned with. The 1974 census did report that 947 farms were owned by publicly held corporations, but most of these were not giants either. Moreover, even if the 947 were entirely in the hands of Safeway or the A & P, it is doubtful that a real threat to competition would result. The 947 corporate operations accounted for 5.7 million acres, $2.5 billion in value of land and buildings, and $2.7 billion in products sold, which amounts to percentages of all U.S. agriculture for these items of 0.5, 0.8, and 2.9, respectively.[13]

The idea that corporations in agriculture are a threat to the public interest seems often to be coupled with a broader thesis that the dominance of the profit motive in and of itself tends to lead to an agricultural and food economy that is not conducted in the public interest. *Farming for Profit in a Hungry World* undoubtedly presents an accurate short description of the springs of action in a world in which food is a scarce good.[14] Harder to come to grips with is the book's insinuation that this is a reprehensible state of affairs. With respect to policy, in any case, it would be unjust and probably counter to the public interest if the government were intentionally to undertake actions that would reduce or penalize profit-seeking in agriculture.

It is true, however, that the average size of farms is increasing rapidly as the number of farms declines. In 1950 there were 5.6 million farms; in 1978, 2.6 million. Between these years the average size of farms increased from 205 to about 400 acres. The farm population has been declining at a trend annual rate of about 3 percent for the past three decades, and continues to do so at about that pace.

A 3 percent rate of decline leads to a halving of the farm

population each generation. This creates substantial adjustment problems for the rural nonfarm population that supplies goods and services to farm people. Together with the substantial trend decline in the real prices of transportation and communications services, it has put an end to a great many small towns and has created hardships in many others. Moreover, the people who are displaced from farms and small towns tend to migrate to cities, where their presence creates another set of social problems. And even in the farm sector itself, while the remaining farmers gain economically from the adjustment process, the process could eventually result in the demise of the family farm and the cultural values that it embodies.

These developments constitute real social costs that our economic calculations of social costs do not take into account. But how can we possibly bring them into a policy-relevant accounting? Some value, implicitly or explicitly, must be assigned to them. The difficulty of making rational judgments in these matters is best illustrated with reference to the idea that off-the-farm migrants are an economic and social liability to the places that receive them. The arguments for this position are similar to those against Indochinese refugees or against illegal aliens from Mexico or indeed against immigration in general. While there are many individual gainers and losers, the relevant question from the point of view of the receiving community is whether the immigrants produce, economically and culturally, more or less than they consume or use up. I believe that we do not have evidence that past immigration into the United States, or from one part of the United States to another, has caused net social costs that are large enough to warrant policies to restrict the movement of people. The argument could be equally well made that the movement of people from rural to urban areas has benefited *both* sectors. Both rural and urban areas have had their problems, but to attribute these problems to movements of the population may confuse symptoms with causes. Costs of adjusting to change constitute a real and important factor in the farm population, in the rural nonfarm community, and in urban areas, but these costs are transitory. The problems relieved by adjustment would, in its absence, be permanent.

There is some evidence that after adjustment has been made to the decline in population in rural areas, the remaining rural residents, both farm and nonfarm, are better off than if the adjustment had not been made.[15] The reason is that the earlier farm problem consisted essentially of an overpopulation of rural people. In short, it seems unwise to attempt a farm-commodity policy that is aimed at social engineering in maintaining a larger number of smaller farms. Let the farm programs stand or fall on their direct economic merits.

Apart from the wisdom of discouraging the growth of farm size, there is a question of how farm programs could in fact promote small-scale family farming. Of course, Congress habitually claims that its programs are aimed at helping, more usually at "saving," the family farm. Certainly acreage allotments, as in the case of tobacco, hinder the expansion of enterprises, and high support prices can keep marginal, high-cost producers from going out of business. On the other hand, some economists argue that current programs (which base benefits on the output that is actually produced) encourage the expansion of the larger farm operations still more. Price supports probably do not have a strong effect either way.

It is obvious that farm programs are not necessary to achieve the demise of family farming. The area in which the trend has gone furthest, the chicken business, has perhaps the least government support of any major commodity. Tax policies, the same ones that result in the low effective rate of taxation of farm income that was mentioned in chapter 1, are probably more important in encouraging large-scale, capital-intensive farming. On the other hand, in the regulatory area, exemptions for smaller operations give them some competitive advantage. For example, as of 1978, the Fair Labor Standards Act mandated the same legal minimum wage in agriculture as in the nonfarm sector. But employers who hire less than five hundred days of work by farm laborers during the peak calendar quarter (roughly seven full-time workers) are exempt. This and other regulatory innovations, together with the savings in the associated paperwork, involving farmworkers provide a real comparative benefit to the self-employed owner-operator. However, they also provide real incentives to the use of bigger and bigger machinery, so that a farmer can expand his scale of operation without getting into labor-management problems.

Thus, the regulatory programs (and the governmental support of research) may be more pertinent to "structural" issues in agriculture than the farm-commodity programs are. This does not mean, however, that before the government took relatively recent action in these areas we consistently pursued the Jeffersonian ideal of small, family farming. It was not done even in Jefferson's day. John Brewster points out that "early land policies were distinguished by extreme inequality of opportunities for acquisition of public land. The smallest unit offered for sale by the Land Act of 1796 was 640 acres . . . and the total price had to be paid within a year."[16] In 1980, Congress and the courts acted to lift acreage restrictions on federally subsidized irrigation projects that had been imposed by the Reclamation Act of 1902. Thus they showed their continuity in spirit with the Founding Fathers. These structural issues go beyond the scope of this book. Nonetheless, while the broader social and cultural issues that a purely economic conception of social welfare leaves out are important, I do not believe that incorporation of them would seriously modify the earlier judgments concerning farm programs and the public interest.

The existence of serious problems with market performance in agriculture, however, does suggest that some sort of government action might promote the public interest. Instead of recommending the replacement of current programs by laissez faire, perhaps we should recommend new or improved programs. The following chapter explores the potential for promoting the public interest along these lines.

6

What Is to Be Done?

In examining the potential benefits and costs of governmental action, two very different approaches may be taken. The first I will call the "policy-analysis" view. It attempts to identify public solutions to problems that arise in the absence of governmental action. Its prime question is: What is the problem? The benefits of governmental action result from the problem's being solved. If the policy analyst can devise a governmental program to solve the problem and if the estimated benefits exceed the costs, he recommends that it be put into action. The second approach I will call the "political-economy" view. It attempts first to understand the economic behavior of governmental entities and then to assess the benefits and costs of the predicted behavior.

Both approaches are necessary in developing sensible recommendations with regard to policy. In agriculture, the policy-analysis approach is essentially a critique of the market. The political-economy view is essentially a critique of government. The preceding chapter reflects the preoccupation of agricultural economists with the policy-analysis view. Having identified the market's deficiencies, one shows how governmental action can remedy the situation. However, this is the point at which the political-economy approach must be considered. Unfortunately, our theories of governmental behavior are not as well developed as are our theories of market behavior. Even so, in the governing of agriculture, we have a long and involved history of governmental behavior to learn from. This experience is relevant to any assessing of the potential for public remedies in each of the five areas where chapter 5 suggested that the market performance was deficient.

STABILIZATION POLICY

The prime area for governmental activity in the public interest is stabilization policy. The best argument for governmental price stabilization—for example, by means of CCC stocks—is that the stocks provide insurance against the rare but extreme shortfalls that would otherwise force large adjustment costs upon the economy. In retrospect, there is no denying that agricultural prices have at times fluctuated needlessly. Yet, historical experience gives us no good reason to suppose that governmental intervention would have succeeded where private action failed.

For example, it was argued for years that while the old sugar program raised U.S. prices above world prices most of the time, the program provided insurance against fluctuations on the high side. Yet, when world supplies shrank and prices exploded in 1974, the U.S. price went right up with world prices. The insurance was worthless when we really needed it. Likewise, in 1977/78, a new sugar program was implemented which boosted returns considerably to producers by holding the U.S. price at perhaps twice the world price by using import controls and stock accumulation. Yet, U.S. consumers paid no less than the world price in 1979/80, when world sugar prices rose. Similarly, the International Coffee Agreement, which the United States joined in 1975, was advertised as a stabilizing device; yet, when the Brazilian frost struck in July of that year, it could do nothing to prevent two years of high prices.

Even in the grain market, where CCC stocks had helped in smoothing out relatively minor fluctuations in prices during the 1960s, they were of little use when stabilization was really needed in the mid 1970s. Indeed, in this episode it seems likely that *government was an important agent of instability:* first through subsidizing wheat exports in 1972 and by not moving quickly enough to dismantle production restraints, then by selling off stocks too quickly in 1973 and 1974 (it was not private speculators but our own CCC that mishandled this), then by attempts to redress the error via export controls during the period 1973 to 1975, and finally by encouraging farmers through 1976 to believe that a new era of high prices and prosperity had dawned, thus promoting the classic cobweb cycle of overproduction in 1977.[1] Moreover, the actions of foreign governments worked to exacerbate the instability of U.S. prices, by attempting to isolate their domestic markets from world conditions of supply and demand.

The reasons for the government's failure in the area of stabilization were not that Congress and the executive branch had bad luck or were incompetent. Just as with market failure, governmental failure arises from the failure, not of individuals, but of institutional arrangements.

The rules for a stabilization scheme—acquisition prices, release prices, size of stocks, storage subsidies—are the outcome of a political decision process. It is difficult to observe the political scene in any detail without becoming skeptical about complex economic justifications of governmental intervention in agriculture or of congressional statements concerning the goals of farm programs which are expressed in broad social and economic terms. When we get to the bottom of it, the ends of politics are the ends of individuals as they act in the political arena—the same kinds of private and personal goals that are taken for granted in the private sector.

There are two ways to answer such a question as Why do we have farm programs? One can give the reasons, or one can give the real reasons. The reasons are to be found either in the preambles to agricultural legislation, which explain what the Congress intends to accomplish, or in textbook discussions of the goals of farm policy. The real reasons have to do with political muscle. Past "stabilization" programs have been more effective in holding market prices up during periods of surplus than in holding prices down during periods of shortage, because that is exactly what they were designed to do. The determinants of this design had very little to do with the public interest in terms of minimizing net social costs. The design had a lot to do with the particular interests of those who would gain most from governmental intervention, the commodity producers. In designing new public policies of stabilization, why should we expect a different result?

I do not wish to oversimplify the lessons that we have learned from the political-economy approach to governmental action in agriculture. There are plenty of genuinely public-spirited people in Washington, D.C. Even when narrower politics rule, the constant shifting of special interests in political decision-making is complex, and political forces do not all pull in the same direction. Why are producers of certain commodities protected while others are not? Why is the farm program budget as big as it is? Why do some programs work through supply control, while others work through government payments? Though the political answers given to such questions are sometimes plausible, they are almost always scientifically primitive. This is not to dispute the general value of recent work by economists which explains political behavior as a continuation of the search for individual wealth by means of collective actions. This line of approach is useful chiefly as an antidote to the view that seeks to defend the social value of farm programs along the following lines: "The Congress of the United States reaffirmed year after year the need for price and income programs to protect and support farm incomes. This it did after reviewing the consequences of the programs each year and evaluating the costs of the programs."[2] What was reaffirmed year after year was that Congress desired such programs, not that there was the "need" for them from the point of view of the public interest.

What sort of governmental action in the area of stabilization holds the greatest promise of dealing with agriculture's problems of instability, given the pitfalls of political decision? First of all, it should be pointed out that the markets are not completely helpless in the face of uncertainty and instability. Individual farmers have recourse to several well-established methods of risk management that can reduce the variability of income. These may be preferable to governmental stabilization of market prices. They include diversification of enterprises, risk-sharing input contracts such as crop-sharing land-rental arrangements, risk-sharing output contracts with middlemen, forward sales, commodity storage, and savings accounts. These all involve costs, however.

Hedging by means of future markets is perhaps the most frequently suggested means of stabilizing farm income through a decentralized market mechanism. It works like this: Suppose a farmer in central Illinois is deciding in February about his production plans for the following year. He can grow either corn or soybeans. Which crop is chosen depends on the price that he can expect for his production. In February, he reads in the newspaper that next fall's futures contracts are selling for $8.00 per bushel for soybeans and $3.10 per bushel for corn. He knows that these Chicago prices usually mean a price on his farm of about 40 cents (the "basis") lower—that is, $7.60 and $2.70. At these expected prices, he would prefer to grow soybeans. But by the time next fall comes around, the situation could be altogether different. Soybeans might sell at $9.00, but they also might sell at $5.00. Corn will not fall too much lower, because of the loan program and the farmer-held reserve program. So, the farmer might grow corn, because this would involve less uncertainty.

With futures, however, the farmer can "lock in" a price of (approximately) $7.60 prior to planting. How? By selling November futures in February at $8.00. He is selling "short," selling beans he does not have. He is selling ahead of time the beans that he will grow that summer, a "forward sale." Then when the crop is harvested in October, the farmer will simultaneously sell his crop on the cash market locally and buy the same number of futures contracts that he sold in February. If all turns out as expected, the farmer sells his soybeans in October for $7.60 and buys back his futures at the same $8.00 price that he sold them for. What if the Chicago soybean price should fall to $5.00 in October, which would amount to $4.60 locally? Selling the soybeans at $4.60 would be a catastrophe, except that now the futures contract that he sold for $8.00 will be bought back for $5.00. Thus, the farmer lost $3.00 per bushel from his originally hoped-for soybean price, but he made an exactly offsetting profit of $3.00 per bushel in the futures market.

This is the meaning of being hedged: the cash position and the futures position cancel out. The $7.60 is guaranteed (minus the brokerage charges and the interest on money that was tied up in the futures transaction). Not quite guaranteed, because the 40 cents basis may change (for example, a transportation tie-up might make the local price $7.20 when the Chicago price is $8.00 in October). There is also the problem that the farmer's soybean crop might fail. Then he is no more hedged than a speculating dentist would be who had sold short in February but had never seen a soybean in his life.

What hedging accomplishes is a reduction in the variability of returns. Unhedged, the $8.00 price might mean a $30,000 income to the farmer; the $9.00 price, $50,000; but the $5.00 price, –$30,000. On these grounds, armies of agricultural extension agents, brokerage-house representatives, and bankers have been urging farmers for years to hedge. Many of them do. Many more do not. Why not? Maybe they don't dislike uncertainty as much as we think they do.

Sometimes farmers will say that selling forward is just as much a gamble as taking one's chances on the cash market. If you sell forward at $8.00, and the price rises later to $10.00, you have gambled and lost. In part, this is just a confusion about the meaning of gambling. It's like saying you are gambling by staying home from Las Vegas because you might have won big if you had gone there. This attitude probably explains why farmers don't hedge more: they really don't want a price to be locked in ahead of time. They want insurance against low prices while still being able to take advantage of high prices.

This desire sounds like a pipe dream, but there exist contingent contracts which can provide exactly this protection. They are called "put options." A put option is a contract that gives a person the right, but not the obligation, to sell at a specified price. A farmer may think: "I would like to insure myself against a price of corn below $2.00, because prices this low would put a severe squeeze on my family's standard of living. But apart from that, I want to take my chances to get a really good price." If the producer acquires a $2.00 put option, he sells the corn at the option price if the market price is below $2.00, but simply does not exercise his option if the market price is above $2.00. There is a catch; namely, the producer will have to pay for his put option. The other party to this contingent contract, the option "writer," agrees to accept corn at $2.00 when the market price is $1.50. In an organized market for put options, a $2.00 put option would be expected to sell for about the amount of money that the writer can expect to have to pay out. So, if there is a 20 percent chance of a market price of $1.50, all other possible outcomes being market prices of $2.00 or more, the option would sell for about ($2.00 – $1.50) × .2 = 10 cents per bushel.

A market in put options does not exist in the United States. Why not? Because in its wisdom, Congress banned options-trading in the major agricultural commodities in 1936. Why aren't the grain producers clamoring for a market in put options? They don't need to buy them, because the U.S. government gives them away. The CCC loan program is in fact a put-option program. In 1980 the producer had the right, but not the obligation, to sell corn to the U.S. government at $2.20 per bushel (wheat at $2.50, soybeans at $4.50, and rice at $6.79 per hundred pounds).

The target prices, too, are similar to put options in that the producer's position regarding contingent income under a target-price, deficiency-payment arrangement is analogous to his position if he had hedged by purchasing a put option at the target price. The producer is guaranteed the target price, but he reaps the benefits of higher market prices. The same is true, subject to program requirements outlined earlier, under the commodity programs. Again, the difference is that the income protection is free of charge to farmers under government programs, but they would have to pay for it in an options market.

Indeed, a better accounting of the real costs of current farm programs than calculations of the kind given in chapter 4 is obtained by asking, What would be the market value of options to sell grain at the target price and to deliver grain to the CCC at the loan rate? The target price can be thought of as a put option guaranteeing a price at harvest time, and the loan rate as a further put option guaranteeing, in addition, the minimum price to be received after a period of storage. These options would have value, of course, even in years such as 1979, when unforeseen events during the crop year caused market prices to rise above target prices and loan rates. The value of the target-price option in any year is the sum of the probabilities that prices will be below the target price, each multiplied by the gain per bushel that farmers would realize from the program at that price. The put options not only have value to farmers; they also have corresponding expected costs to taxpayers, even in years when they are not exercised.

A commodity-options market that is functioning well would be an excellent substitute for the income-stabilization features of current farm programs for grains, rice, and cotton. In fact, it would be better, in that each farmer could choose the degree of price insurance that he wanted by purchasing a put option at the appropriate guaranteed price. Of course, the higher the insured price, the more the farmer would have to pay for his put option. If a farmer wanted a complete forward sale, he could sell futures or some other form of forward contract. The key feature of the put option is that it duplicates the kind of protection that farm programs give—a price floor but no limit on profits from high prices.

While options would provide farm-income *stabilization,* they would not provide farm-income *support.* Thus, they would have neither the income-redistribution nor the resource-allocation effects that price-support programs have. This feature would make the option-market approach more nearly congruent with the public interest, but it would not be politically popular among farmers.

There are several objections that might be made if price guarantees for producers were replaced by commodity options. First, there are at present no options markets, and it would take years to implement them. This fact means, of course, that options could not replace the current programs overnight. The immediate step to be taken is to remove the legal prohibition against trading in options that now exists. Indeed, after making options legal, the CFTC (which has jurisdiction over this area under current law) should immediately move to encourage and assist the organized commodity exchanges to develop trading in options.

In the absence of an organized central exchange for options, they could nonetheless play a central role in farm programs. The CCC itself could be the writer of put options. It could draw up contracts that would give farmers the option to sell at, say, the loan rate. It could also write contingent contracts that would be comparable to current target-price protection. These would provide indemnity payments to producers that would be determined in basically the same way as deficiency payments are now—they would provide market-price insurance. The difference would be that instead of giving away the insurance, the USDA would sell it. Farmers could buy these put options (CCC market protection) or price insurance (target-price indemnity payments), for example, at the county ASCS office. If for political reasons it were necessary to subsidize the price insurance, the USDA could price these options at less than their expected value. This is, in fact, exactly the approach that is being proposed for a comprehensive scheme of crop-production insurance to cover natural hazards. All that is proposed here is to handle price insurance in the same way. A side benefit is that the program would make honest options salesmen out of the ASCS employees, who would be relieved of some of their administrative burdens, since incentives to overproduce would be lessened.

Another objection to either futures or options as substitutes for price supports is that these contracts are for relatively short periods, never much more than one year for even the best-developed futures, while producers most desire protection for periods of more than one year. They are investing in extremely expensive capital equipment—in 1980 a combine iteslf could run to $80,000—and they want assurance that their investments will not be a total disaster. To counter this problem, the USDA could offer longer-term options, even if the nation's doctors and lawyers did not come forth to write them in sufficient quantities to make a central market in long-term options viable. Of course, the required payment for a long-term option would be higher than for a short-term option that guaranteed the same price (because there is an increased probability that the market price will fall below the price guarantee, given a longer time period in which it may do so). This kind of option could, in fact, provide better long-term price insurance than current programs do. In the past, real support prices for grains have in fact fallen substantially when longer-term market forces have pushed in this direction. Thus, it is still risky for a farmer to count on government price supports to guarantee a profit from a long-term investment in equipment.

Finally, it may be objected against the commodity-options approach that while it would handle the farmers' problems of variability in income (albeit at a price to them), it would not stabilize actual market prices, and so would not solve the consumers' difficulties in facing variability in prices. This need not be a problem for large-scale consumers, such as grain millers or big feedlots, because they can buy forward via futures. Or, with a commodity-options market, they could buy call options—options to buy at a specified price—which work just the same as put options do, only on the buyer's side. But to stabilize prices for final consumers of a good, to ensure supplies for food-aid purposes, or to dampen prices under extreme shortfalls in order to forestall macroeconomic dislocations, we need actual management of markets—specifically, of stockpiling—not just price insurance.

There is indeed a legitimate role for government intervention in markets for this purpose. But this does not compromise the simultaneous use of a commodity-option scheme. Perhaps the chief risk of "stabilization" schemes is that for political reasons they become income-support schemes and then lead to set-asides or other forms of production control which are wasteful of the nation's resources. If there were a system of price insurance, a stockpiling/stabilization program could focus on the socially useful functions that it ideally can serve. Moreover, commodity options can facilitate the operation of a public-storage regime. Either the CCC's own put options or put options acquired by the USDA on an options market could be used, as the loan rate is today, to acquire grain for storage. And call options could be used to hold grain or to buy for shipment overseas or for whatever other purpose the government might wish to have guaranteed access to grain in times of shortage.

These devices would give the government better control over grain stocks than is now possible under the farmer-held reserve program. In this program, farmers hold grain under CCC loan, subject to call at a trigger price (175 percent of the loan rate for wheat). But the CCC's call option only pertains to the *funds* loaned. The producer may do as he pleases with the actual grain. With call options on the grain itself, the government can ensure the delivery of grain for such purposes as emergency aid, even when supplies are short. In addition, call options would fit in well with an international "grain insurance" scheme as proposed by D. Gale Johnson.[3] The idea would be that the United States would replace its current food-aid programs (except for extraordinary emergency relief) by an offer to poor countries to make up any shortfall in excess of, say, 6 percent of their trend production (which Johnson estimates would require an average annual payment of 5 million metric tons). Call options could be used to acquire grain for this purpose.

The government itself need not be in the storage business in order to promote increased stockpiling and increased price stabilization. Grain insurance or other firm food-aid commitments would effectively monetize the prospective emergency demands of poor people abroad. If we did not wish to rely on this mechanism to encourage private profit-seeking storage and if we wished to provide additional stocks for purposes of domestic macroeconomic stabilization, government-held stocks could be used. The quantities that would be required would probably not exceed the Carter administration's proposed program to hold 200 million bushel (5½ million metric tons) in emergency reserves of wheat. In fact, this seems to me to be the only current program that it would be in the public interest to retain essentially as is. If further incentives to stockpiling are desired, a simple subsidy to private grain storage of, say, 20 cents per bushel per year would probably be the most efficient way to promote added price stability. A simple subsidy is, I believe, preferable to our current complicated system of price triggers and payments under the programs of CCC-loans and farmer-held reserves.[4]

The list of policy proposals in this area is greatly enlarged by consideration of international agreements in grains, particularly wheat. There are many alternatives, none of which are extremely objectionable as long as they are used solely for stabilization purposes and not for farm-income support. All of them would be facilitated by the existence of commodity options.

While market institutions in agriculture cause problems of instability and uncertainty, individuals are often able to ameliorate these problems by themselves in a market context if they are left free to do so. The adjustments may be quite complex, as with commodity options or insurance markets, but they can also be accomplished by individual action. An elementary way to cope with instability is to save money when income is temporarily high, and then un-save it when income is low. This is exactly what farmers tend to do. Indeed, among farmers, the marginal propensity to save is so strikingly high (0.4 to 0.5 compared to less than 0.1 for nonfarm families) that observation of this propensity in early U.S. budget studies was one of the main bits of evidence that led to the permanent-income hypothesis.[5] Thus, the study of the uniqueness of agriculture can be valuable, not because farmers are shown to be subject to different laws of economics, but because behavior that is common to all can be observed in agriculture to an intensified degree.

OTHER SPECIAL CHARACTERISTICS OF AGRICULTURE

Critics of governmental intervention have themselves often been criticized for being ignorant of the complexities of the real world which have made the intervention necessary; yet, at the same time, government policies have often caused problems precisely because they have run afoul of the complexities of agriculture. Thus, the wheat program has been oversimplified by applying the same target price to what really amount to different crops, grown under regionally varying circumstances that cannot possibly be taken adequately into account. The cotton program has formerly treated skip-row planting (leaving fallow rows between rows of cotton) as diverted land both in the Southeast, where it makes some sense as diversion, and the Southwest, where it is a valuable production practice to conserve soil moisture and is not production control at all. The Agricultural Conservation Programs, more broadly, cannot hope to distinguish production-promoting practices from conservation practices, and we have ended up by paying farmers to create new cropland from forest and swamp while simultaneously paying them to idle cropland that is already productive. With regard to trade problems, tariffs or trade restraints on particular products have had unwanted effects on trade in related finished or semifinished products (such as sugar) and on materials used in a product (textiles and cotton) and on close substitutes for the product (raw and processed meats). (For other examples, see chapter 2.) The overall result is ever-increasing complication of programs to plug "loopholes" that keep appearing.

Even in the area of the general rationale for farm policies, the government has been blind to the special characteristics of agriculture. The idea of "cost of production" as a basic guideline to what a minimum price should be does not make sense in the presence of price instability. If the mean price were to be equal to the cost of production, however defined, then the annual market price would be below the cost of production roughly half the time. Thus, a support price below the mean price can cause considerable change in the price expectations of farmers. Yet, officials in Washington have been known to argue that target prices at current levels will not induce increased production because they are below the cost of production.

In justifying set-asides as a means of supply management, it has been argued, in criticism of the market, that the complexities of producers' psychological and economic behavior in a dynamic situation with long lags require governmental inducements to make the proper adjustments. However, the government's use of set-asides is not purely a matter of economic rationality in supply management either. The 1978 decision to pay producers to grow less wheat and feed grains did not look good in 1979, when prices rose to the "shortage" levels calling for release of grain-reserve stocks. This initial and presumably well-thought-out use of set-asides under the 1977-act programs raises doubts as to whether the unique, dynamic complexities of agriculture are something that we ought to look to the U.S. government to deal with.

In seeking the reasons for this state of affairs, it is useful to look a little further into the political economy of farm policy. During the past decade, economists have become attached to the phrase "public choice" in their thinking about political action. One can, with some stretching of the term, treat voting by the public or voting in Congress as acts of decision by the citizenry. The range of governmental action is narrowed, but not usually determined, by these decisions. Many of the real "public choices" in farm programs, as in other areas, are made within the executive branch. But it is no easy matter to determine how and by whom a particular choice has been made. The president is the ultimate decision maker, but in specific areas, such as agriculture, the decisions in many cases are made de facto by subordinates. Moreover, governmental action is not determined by any single decision.

The public decision process can be thought of as follows. Initially, there is a wide range of possible programs for, say, intervention in the wheat market. The probability distribution of outcomes is narrowed by the voters' choice of congressional representatives. It is further narrowed in the give and take within the Congress that results in the passage of an actual piece of legislation. The executive branch then, at the technical staff level, narrows the range of concrete program options; interagency disputes are passed up to higher subcabinet or cabinet-level decision points; and those that are both irreconcilable and important enough go on up to the president. Then, after a concrete choice has been made, the courts or the executive branch or Congress can respond to citizens' reaction by making further revisions. Some courses of governmental action are ruled out at each stage. But in this process of narrowing down, the participants at each stage of the process are aware of the other stages: for example, people on the White House staff know what the president's general views are. In short, it is often quite impossible to specify a locus of public choice more precise than "the U.S. Government in Washington."

The pitfalls that stand in the way of wise decision-making come mainly at the highest levels of the legislative and executive branches. Consider first the Congress. The key problem is that with so many weighty matters to worry about—such as the nation's defense, the macroeconomy, and dealing with their staffs and constituents—there is simply not enough time for Congress as a whole to become properly informed about something so complex, yet so relatively unimportant, as agricultural policy. Thus, policy in this area tends to be left to those congressmen to whom it is important, in the relevant committees. But, of course, these committee members are selective, not general, in the interests that they represent. And they, too, are too hard-pressed to become astute students of the complexities of each commodity and each program over which they have jurisdiction. They naturally turn for advice to the most congenial lobbyists and trade experts who express interest in agricultural legislation and to the staff of the relevant agency of the executive branch, usually to the USDA. These experts and interested parties have interests that are even narrower than those of the committee members. They are much more concerned about an appropriate division of the U.S. economic pie, from their own point of view, than with lofty goals of curing ills from the point of view of the public interest. Congress thus becomes suitable for a metaphor à la Clausewitz—it is a continuation of the struggle for wealth by other means, a tool by which organized groups of citizens attempt either to increase their slice of the economic pie or to prevent others from decreasing it. Thus, while we may give an economic-theory rationale for Congress to ban futures trading in potatoes, for example, the reason that this becomes a legislative matter is the strong feeling of an organized group that they would do better financially without the futures market.

Much farm legislation has granted wide discretionary powers to the executive branch. Not only has Congress passed many decisions on to the president or to the secretary of agriculture, but the ones that have been passed on have typically been the difficult decisions. One of the things that Congress does most of, after cutting pies, is passing bucks. Congress often wants to "do something," but it does not know how to; or a division within Congress makes buck-passing a useful compromise. For example, Congress has agreed to provide the U.S. beef industry with some protection against imports, but it has recognized that this protection, under at least some circumstances, might not be in the national interest. Being unable to specify what these circumstances are, Congress, in its legislation regarding imports of meat, empowers the president to make the determination. The president has authority to suspend import quotas if he finds it to be in the national interest to do so.

In making such judgments, any administration, in weighing the pros and cons, depends heavily on estimates of how much consumers will gain and producers will lose, with political weights attached. But equally important are judgments about the broader economics of intervention. Proponents of the regulating of imports say that if imports are not restrained during periods of low U.S. prices, cattlemen will liquidate their herds, and prices will soon rise higher than they otherwise would have. The fact that there is a quasi-cyclical liquidation and expansion of the aggregate cattle herd, which economists are not able adequately to explain or forecast, gives a superficial plausibility to such arguments, or at least it precludes an empirical refutation of them. The lack of solid knowledge means that when a committee of government economists, each representing different interests, meets to draft a description of the options for the president's consideration, the resulting collective economic analysis may say: "Reducing imports will over the long term either increase or decrease U.S. meat prices (or else it will leave them unchanged)." In the end, one view or another will prevail, at least partially, and thus the resulting treatment will usually not be so two-handed as to be useless. But typically, a strong case for a single option on economic grounds will not be presented.

The source of trouble is what seems to be a general problem—namely, that the executive branch of the U.S. government is simply asked to do too much. Legislation that clearly spells out objectives and procedures to attain politically decided ends is not so problematical. A cabinet agency can, in general, be expected to carry out its instructions if it knows what they are. The problem lies in the vagueness and the broad discretionary powers that are granted in agricultural/food legislation. This is a two-edged sword. Given some of the bills that have been passed by Congress in pursuit of special-interest votes, such as the Meat Import Act, it is in the public interest that the president have discretionary powers. But this idea too easily becomes a mechanism by which Congress can ostensibly provide special interests with what they desire without really doing so—a temptation to force ever-more-difficult administrative tasks in ever-greater volume upon the White House. Indeed, the existing farm legislation, taken together, empowers the executive branch virtually to be the manager-in-chief of U.S. agriculture.

Moreover, it turns out in practice that the executive branch's policy-making in agriculture is not as well tuned to the general interest as the broad-based nature of the presidency might suggest. There is, in fact, considerable ambiguity as to who actually makes decisions regarding farm programs. The center of events is, of course, the presidency. Even decisions that have explicitly been delegated by Congress to the secretary of agriculture often involve White House decision-making. The secretary is, after all, an employee of the president, and the secretary's decisions have to fit in with the president's program as a whole. At the same time, it goes without saying that presidential attention must be directed first and foremost at the jobs that are most vital to the nation that will not be done unless the president does them. Thus, issues of international affairs and national defense come to the fore. Within the realm of economics, the first order of priority is necessarily the macroeconomic issues—issues concerning the nation's rate of inflation, unemployment, and economic growth.

Where do agricultural and food issues fit into the executive branch's decision-making? There are two countervailing forces. First, agriculture in and of itself is becoming less and less important, both economically and politically. But second, agriculture is becoming more and more integrated with other economic and political areas and is therefore a matter of broader concern. For example, food and agriculture are of greater importance in international affairs, and hence of greater interest to the State Department, than once was the case. The Treasury Department, the Justice Department, the Council on Wage and Price Stability, and the Labor Department have also seen reasons to want to increase their input into agricultural policy-making.

The net result of both forces is that issues of agricultural policy are delegated to the subpresidential, or even to the subcabinet, level for the main analysis of them and for all but the most important and difficult decision-making. But the Department of Agriculture works out less and less well as the appropriate subpresidential agency for dealing with issues, because of the multiplicity of interests involved. This tends to make the White House staff a focal point for executive-branch discussion of food and agricultural policy.

Nonetheless, it is in the USDA that the experts on the many technical matters involved in decision-making reside, and it is there that all but the most sensitive matters are resolved. Here again the interests that are represented are inevitably narrowed. Indeed, one finds that de facto policy decisions are being made by agencies within the USDA over which the secretary of agriculture has only limited control. At times, the Agricultural Stabilization and Conservation Service has had its own policy stance. During World War II, the secretary of agriculture was working for increases in farm output as part of the war effort, while the predecessor of the ASCS was working to hold output down in fear of postwar surpluses.

As an example of largely autonomous policy-making within the USDA, there has recently been a special impetus given to expansion of cooperatives in the grain-storage business. The secretary of agriculture, in 1977, ruled that cooperatives could put grain under CCC loan for their members, an attractive option that was not open to commercial grain elevators. Thus, cooperatives could offer more attractive terms than commercial elevators could to their customers. Similar provisions in cotton have helped cooperatives almost completely to replace other commercial enterprises in the cotton-storage business. This decision—negligible in the big picture of the U.S. economy, but potentially fundamental in grain-marketing—was taken without serious analysis or discussion outside the USDA.

The difficulties in attaining wise presidential leadership in agricultural policy are serious from a purely analytic and managerial viewpoint, but perhaps what is even more problematic is the role of politics in the White House. When it has something forceful to say, economic analysis rules in the White House decisions more often than in Congress. But presidents, too, are susceptible to the same short-run, special-interest political pressures as congressmen are. The effect is notable, particularly in election years. The period immediately prior to the 1976 election offers two telling examples of the role of political calculation: the decision to raise the loan rate on wheat from $1.50 to $2.25 per bushel and the decision to triple the tariff on imported sugar from 0.625 to 1.875 cents per pound. The decisions were basically political ones and would not have been made had the election not loomed so large in White House thinking. We may expect some of the same in every presidential election year.

In sum, I do not doubt that the market as an institution has often malfunctioned in the face of the uncertainty and complexity of agriculture. But I believe that a study of the U.S. government's past and current actions in farm-commodity markets and of the political economy of the public decision structure that gives rise to these actions makes it excessively optimistic to look to the government to remedy the market's deficiencies.

PUBLIC-INTEREST REGULATION OF THE FOOD SECTOR

Perhaps we can be rescued from complete pessimism about promoting the public interest through the governing of agriculture by calling the middlemen back to the dock. We earlier concluded that antitrust action seemed to be in order. Intervention by the Federal Trade Commission (FTC) and the Department of Justice to promote competition in food retailing and food processing is a good example of how legislation can be enacted and implemented which does not serve the special interests of those whom it affects most directly.

Yet, the story here is not so encouraging either. Consider the moves against Iowa Beef Packers, Inc. (IBP) by the Packers and Stockyards Administration and by the Justice Department. Expansion by IBP is a prime factor in the rapidly increasing concentration in beef packing. If anybody will ever have monopsony power over producers of agricultural products, IBP will. Yet, the main reason that IBP has been able to expand is that its lower costs have enabled it to undercut sellers of carcass beef while simultaneously offering better deals to cattlemen than competing packers can afford to. Suppose it turns out that because of economies of scale or for other reasons, IBP can offer farmers better terms while obtaining monopsony rents than are being offered under the present more-competitive market structure. Where, then, is the social gain in restricting the expansion of IBP?

The difficulty in regulating competition is that where the political will can be mobilized, the economic arguments often begin to look weak. The resulting conflicts among politically mobilized experts tend to be self-canceling, and we cannot even be sure that this is an undesirable outcome. For while we can agree that the markets are imperfect, we can't agree about how to make them function more nearly in the public interest.

This situation is further complicated by the burgeoning institutional innovation that is taking place in the marketing of farm products. A desire for improved risk-management by specialized, capital-intensive, and highly leveraged producers, as well as by their bankers, coupled with a desire by processors to ensure access to supplies, has led to many new forms of pricing and contracting for farm products. Although questions have been raised about the desirability of these changes, there is no concensus on what would be the economically appropriate regulatory measures, if any.

THE PURSUIT OF BROADER SOCIAL GOALS

The declining number and the increasing size of farms indicate several problem areas where governmental action might promote the public interest. These could be broadly characterized as "externalities," costs imposed by market outcomes upon parties that are not directly involved, and "public goods," commodities or services that can be cheaply appropriated by many, and so provide insufficient incentive for private profit-seeking enterprises to produce them.

Adjustment of past dislocations has imposed heavy costs on rural people. The appropriate provision of public goods in several areas could help in coping with future adjustments.

Information. Both privately and socially wasteful decisions regarding the allocation of resources can be minimized when pertinent market intelligence and outlook information are available. The fact that market statistics and price forecasts, once they have been released, are accessible to everyone suggests a lack of incentive for investment in information-generating activities in a free-market setting. There are, of course, private businesses in the field—such as Doane's Agricultural Services, Data Resources, Chase Econometrics, and Wharton Econometric Forecasting Associates—but further public investment in dissemination of information is likely to have benefits that exceed the costs. Even the statistical activities that budget cutters are most anxious to eliminate, such as statistics about the mink industry or the survey of white corn, could well have net social benefits. The costs of avoidable errors in production, even for a small-scale industry, can be great relative to the costs of a survey. Although it might be preferable to have the industry, rather than the taxpayers, finance the surveys that the industry wants to have undertaken, the transfers in this case are more likely to be part of a positive-sum game (i.e., the gains of the winners will exceed the losses of the losers) than the transfers under most agricultural programs.

Perhaps the single most important source of market information for farmers is publicly quoted prices of transactions in centralized cash and future markets. These give a valuable indication of current estimates of the balance between supply and demand as related to past history and the outlook for the future; yet, their traditional mistrust of middlemen has led farmers to acquiesce in and even agitate for the suppression of this information. The USDA is explicitly prohibited by law from forecasting the price of cotton. Futures markets have often come under attack, and from time to time they have even been banned. In August, 1979, after a two-year hiatus, the New York Coffee and Sugar Exchange resumed its quotation of a New York price for raw sugar. The hiatus resulted from a Department of Justice suit objecting to the way in which the quotation was arrived at. These restrictions can only harm the position of the decision maker on an ordinary-sized farm relative to larger businesses, which can better afford private information systems.

An additional argument for legalizing commodity options markets is that the quoted prices would provide valuable information on the range of uncertainty in market expectations. For example, if an option to buy soybeans next November at $10.00 per bushel sells for 50 cents per bushel, while at the same time November futures are selling for $8.00 per bushel, the information can be used to provide an indication of the chances that price will rise above $10.00 by November. With market quotations on a series of put options and call options at different soybean prices, inferences can be drawn about the whole probability distribution of expected future prices. This information could be quite helpful to producers in considering their production decisions and their marketing alternatives under uncertainty.

Research. Scientific and technical knowledge can be viewed as information, and it has similar characteristics as a public good. Governmental activity in agricultural research and the dissemination of research results to farmers through the county and state offices of the Extension Service (not to be confused with the ASCS) long antedated farm-commodity programs. The Land Grant Universities, in which most such research takes place, were established by the Morrill Act of 1862.

Public agricultural research has been criticized. At present, there is even a suit seeking to halt such activity at the University of California, on grounds that it favors large producers—for example, by aiding in the replacement of hand labor by the mechanized harvesting of vegetables. It is true that particular client-consultant research could and probably should be done by private consulting firms. Where results are patentable, much agricultural research is in fact being done in the private sector, including research on hybrid corn, improvements in tractors, and the development of chemical pesticides. Benefit/cost studies of public research in agriculture have been conducted to an extent and of a quality that is far above any analytical results available on other public undertakings in agriculture. These studies indicate that the social rate of return on this research has been very high. With respect to the distributional effects of agricultural research, while reductions in costs of production have created economic gains for low-cost producers, the main beneficiaries appear to have been consumers, through lower prices. But whoever gains, the public interest appears, by and large, to have been served, in that the aggregate gains tend to exceed the costs.

Regulation. Governmental regulation of markets can serve the public interest when externalities result in net social losses in unregulated markets. EPA regulation of environmental damage from feedlots and pesticides can be justified in this manner. OSHA and other worker-protection regulation and the regulation of food additives, grading, and labeling by the FDA and the USDA may also be put under the externality umbrella, although the arguments here are somewhat strained. It may be said that an external cost may result from ill health among workers or consumers, in that the general public pays part of the bills through the public health-care system. The aspect of this situation that is more directly relevant to policy is the public health-care systems; that is, it might be better to correct the situation of, say, smoking-related illness by shifting the health costs to smokers rather than by regulating the cigarette market.

Labeling and grading may be thought of as investments in information like those discussed in the preceding section, but there are two important differences. The regulations often consist of having a governmental agency tell a private business what, or what kinds of things, it must say about a product. And when the government itself provides the information, as through "nutrition education," there is often great controversy as to what the content of the message should be. Throughout the regulatory field, the issues are ones of benefits relative to costs, and solid information here is scarce for almost every regulatory area.

One general reason for skepticism about the chances for furthering the public interest through regulation may be gleaned from the preceding chapters. After all, what is described there is basically the federal regulation of commodity markets. This sustained and detailed regulatory effort—both when it is intended to aid producers and when it is intended to aid consumers (through price ceilings or export controls)—has more often than not done just the opposite of promoting the public interest. Why should we expect systematically better results in the newer regulatory areas? One idea is that special interests are the force behind the regulation of commodity markets, but not in the newer areas. This is surely a false dichotomy. Even though some lobbyists for interest groups in the newer areas have had the foresight to affix the label "public interest" to their organizations, this no more settles the matter than do the assurances of producer groups that price floors are in everyone's interest.

Some of the best prospects for fostering the public interest in agriculture and food come in deregulation. For example, ICC trucking regulations have increased the cost of shipping agricultural products by restricting the use of "backhauls" (return trips) for carrying agricultural products. At the same time, ICC rates for rail freight underprice the shipment of farm products, notably grains. This is a two-edged sword: it makes it cheaper for farmers to ship grain by rail, but it also makes it more difficult for farmers to gain access to railroad cars, since the railroads will prefer to allocate cars and capital to goods that provide higher returns. It also induces the railroads to invest less in the covered hopper cars that are best suited for grain storage than a socially optimal allocation of resources would call for. Thus, we read each year about grain that is in danger of rotting because it must be piled in the streets because the grain elevators (temporary storage and assembly points) are full because railroad cars are not available for shipping the grain. This does not happen because of bumper crops or poor planning or market failure. It is a rational response to economic incentives, which has been created by regulation. (It should be said, however, that storing of some grain for a short time in uncovered piles in dry climates would be economically rational even under an optimal policy regime. It does not pay to keep a sufficient stock of cars or elevator space to take care of all grain the moment it is harvested, when these will afterward remain empty for eleven months of the year.)

Stability in the socioeconomic environment. The most basic public good for agriculture is the provision of an environment in which farmers, middlemen, and consumers can plan and act with confidence that the "rules of the game" and the essential infrastructure for economic activity will be relatively stable. The importance of such stability should be evident if one considers the damage that is inflicted by the phenomenon of inflation. Commercial farmers have been substantial gainers from recent accelerations in the rate of inflation, but at the same time they have been put in a hard-to-manage cash-flow squeeze and are subject to an extraordinary degree of risk. Perhaps the most important measures that could be taken to aid commercial farmers and to promote the public interest would be macroeconomic policies to keep inflation under control.

Establishing justice. By an effort to secure justice, I mean not simply a framework of law but distributive justice in market relations: the idea that farmers should receive fair prices. This is a politically potent, if imprecisely defined, idea among the public. The special weight of special interests notwithstanding, some basic elements of farm policy have usually had widespread support, which has been fostered by a belief that farmers—many of whom are very poor, near bankruptcy, exploited by middlemen, or pressed by large corporations—simply cannot, through the marketplace, obtain fairness in prices or in returns on their effort.

What is the source of this notion? Perhaps it is the inveterate pessimism of farmers and their spokesmen. But I believe that a more important source for this perception of farm poverty may be that too many of us get our data from television newsclips at times when farm issues are in the news: for example, when the AAM demonstrated in Washington in 1978 and 1979. At these times, one could hear statements such as the following from the United States Senate:

> Senator Talmadge: The entire agricultural economy of the nation is caught up in a crisis of major proportion that has frightening implications far beyond the farm.
>
> Unless bold steps are taken, and taken quickly, this crisis will spill over to the entire economy and will have dire consequences on the lives and security of millions of Americans in the nonfarm sector.
>
> Senator Allen: Certainly the hour is growing late if we are to save the farmer from economic collapse.
>
> Senator Melcher: We have got to have higher loan rates. We have got to close the loopholes on beef imports. We have got to help cotton, protect wool, and we need to protect sugar growers.[6]

Most Americans would probably agree that governmental aid to people during crisis situations is justified—indeed, that it is morally repugnant to refuse such aid. Therefore, the irresponsible falsity of the preceding statements is serious.

It is possible that even though farmers on a commercial scale as a group are generally not on the verge of bankruptcy, nor are they even poor or near-poor, they still are not receiving fair prices or incomes. Fairness here is roughly equivalent to equity, which does not imply equality. The full posttax incomes of commercial farmers are, on the average, well above the average in the nonfarm population. Nonetheless, it could be claimed that—considering the risks that commercial farmers face, the sums of capital that they have invested, and how hard they work—they are still underpaid.

These considerations lead to attempts to define fair (just, equitable) prices, returns, or incomes. Some such conception lies behind the appeal of "parity" prices. The idea of parity, which practically resolves to indexation, has become obsolete, because the economic relationships that provide plausible links for indexation (say, of the price of corn to an input price index) change too much. For example, the index of prices paid by farmers increased about 77 percent between 1965 and 1975, so that a parity price per bushel of corn, indexed to input prices, would also increase 77 percent. But U.S. corn yields increased from 74 to 101 bushels per acre during this period, so that the cost per bushel must have risen much less than 77 percent. A 77-percent policy-imposed rise in price would have generated unwarranted windfall profits to producers.

With the intention of overcoming this difficulty, current farm programs, with the notable exception of the dairy one, have moved away from parity to cost-of-production estimates as a basis for fair prices. However, there is an impressive list of difficulties with this approach. (1) The estimated costs vary enormously from one region to another and from one farm to another within a region. (2) Some items of cost are almost impossible to attribute to the production of a particular commodity: for example, the cost of the farmer's pickup truck. (3) Some items—notably land—generate returns other than through production of a commodity, and therefore their costs should not be charged entirely to that commodity. (4) Randomness in yield means that costs may vary from year to year, with the result that the supporting of prices at an average cost will inevitably result in average prices being above the average cost. (5) Most fundamentally, at whatever level a price is set, costs will tend to rise to meet it—for example, if the price of wheat were set at $15 per bushel, the prices of scarce wheat-growing resources would be bid up enough to make the cost of production $15.

Congress has mandated commodity-by-commodity surveys of U.S. costs of production, which have been carried out since 1974. As an aid to scientific policy-making, they must be written off as a loss. The conceptual difficulties are such that the estimates can be used to justify almost anything. For example, when President Carter proposed his first wheat target prices in 1977, he explained their levels in terms of having them cover nonland costs plus 1.5 percent of land costs. After Congress had boosted the target prices in the bill that was actually enacted, the level was justified as covering nonland costs plus 3.5 percent of land costs. Six months later, when the American Agriculture Movement brought pressure to bear, the wheat target price was raised another 10 percent; so the cost-of-production rationale had to change again.

What, then, should be our guide to the appropriate support prices? My view is that, in the absence of programs, there is no better guide than estimated market prices. This idea is not an entirely new departure in U.S. farm legislation. The Food and Agriculture Act of 1977 prescribed that the secretary of agriculture should determine the desired output for the major grain crops. Moreover, it told him, roughly, how to make the determination:

> The national program acreage for feed grains shall be the number of harvested acres the Secretary determines . . . will produce the quantity (less imports) that the Secretary estimates will be utilized domestically and for export during the marketing year for such crop. If the Secretary determines that the carryover stocks of feed grains are excessive or an increase in stocks is needed to assure desirable carryover, the Secretary may adjust the national program acreage by the amount the Secretary determines will accomplish the desired increase or decrease in carryover stocks. [U.S. Congress, 91 Stat. 931]

Thus, the secretary is supposed to estimate the quantity that will balance supply and demand, plus desired adjustments in carryover stocks. Given this estimate, a natural candidate for a target price is a level that, it is estimated, will bring forth the desired supply. In a context of long-range planning, one would want a regulated producer price floor that would be consistent with supply/demand balance without any change in stocks. In short, our search for fairness in pricing has led, not by normative argument, but by the absence of a plausible justification for anything else, back to market pricing.

The robustness of the reasoning that leads to market-clearing prices is apparent in the convergence to this position in Soviet planning:

> The system of agricultural prices is economically justified only when, on the one hand, it . . . materially stimulates the steady growth of production in conformity with the planned rate. . . . On the other hand, this system provides for a sufficiently low level of state retail prices . . . to satisfy to the fullest possible extent the people's requirements in food stuffs and articles made from agricultural raw materials.[7]

That is to say, roughly, prices should balance supply and demand. Beyond this it does not seem possible to specify what fair prices should be in any policy-relevant sense—prices which it is the obligation of the government to establish.

INTERNATIONAL AFFAIRS

Given the lack of unregulated markets among U.S. trading partners, there is a natural and useful role for government-to-government negotiation. Probably the most important is negotiation in order to lower trade barriers. The multilateral trade negotiations that were concluded in 1979 did not fundamentally reduce world protectionism, but given the temper of the times, it is perhaps sufficient that they did not sanction increased trade restrictions.[8]

Negotiations for international commodity agreements are not so clearly productive from the point of view of the public interest. These involve internationally agreed-upon price floors and ceilings, usually with management of buffer stocks or coordinated controls on exports in order to defend the price band. Since the early 1960s, the United States has supported the International Coffee Agreement, not as a boon to us as consumers of coffee, but as a way of aiding the coffee-growing countries. The exporting countries are always tempted to use these agreements as price-boosting cartels rather than as purely stabilizing measures, and the United States has been too ready to acquiesce in this approach for cocoa and sugar as well as for coffee.

Bilateral agreements have been reached, in recent years, between the United States and several other countries with reference to trade in grain. The Soviet Union has agreed to buy at least six million metric tons of U.S. grain each year, beginning in 1975, and the United States has agreed to sell to the Soviets up to 8 million tons without consultation, although the permission of the U.S. government is required if they wish to buy more. Informal agreements have been concluded with Japan, Israel, and several other countries. Such agreements may be desired by U.S. trading partners as assurance that they will not be subject to ad hoc suspensions of U.S. exports. Apart from this function, it is not clear that these agreements serve any useful purpose. They may even be counterproductive by reducing confidence in the availability of U.S. supplies among trading partners who have been left outside the agreements and by leading to general expectations that the U.S. government will continue to attempt to manage the grain-export business.

The fundamental problem in the international realm is that while the government is just as likely to fail in the area of negotiating the removal of trade barriers as in other policy areas, here there is no plausible private-market substitute.

CONCLUSION: A PLEA FOR THE MARKET

In reaching conclusions about "what ought to be done," one's values and preferences come to play a major role. Yet, it would be wrong to attribute differences in recommendations on political action solely to differences in values and preferences. Indeed, in the area of values, there is probably quite widespread agreement on many notions: namely, that farming should be an economically viable industry, that conduct and pricing should be in some sense "fair," that food products should not endanger the health of consumers. These and many similar propositions are generally agreed upon as being desirable.

The main disagreements arise in predicting how agricultural commodity and resource markets would operate in the absence of governmental intervention and in predicting what types of intervention the political forces will lead the government to undertake. Where I find myself most often in disagreement with interventionist arguments is not so much in the assessment of the problems that arise in unregulated markets as in the likelihood that even worse economic mischief will arise from political sources when markets are regulated.

I do not deny that there are some areas in which governmental action is the less undesirable course of action, and I would even concede that progress in economic policy is possible. When economic analysis has nothing to offer, political considerations rule; but when economic analysis can provide convincing forecasts, it is sometimes potent. Unfortunately, this almost never happens, except when a policy intervention has been tried and there actually exist "experimental" results to work with. It seems unlikely that we will again have market support prices as high as in the past, and ceilings on food prices are not in the cards. The substantive dissuading force in these cases is not political experience but economic experience. I believe that further analytical progress will eventually lead to more rational farm policy, which will be in the interests of farmers, consumers, and taxpayers jointly. And I believe that these improvements will, by and large, involve extending the role of the market as the pricing and resource-allocating mechanism in agriculture.

The deregulation of farm-commodity markets will be strongly resisted by producer interests, and with good reason. They have purchased land and have made other large capital investments at prices that incorporate substantial program-induced rents in some cases. Nonetheless, producers tend to overstate the chaos and depression that greater market orientation would bring, mainly because they discount the chances for private-market institutions that would assume some of the risk-management services that are now being provided by government. This is why it is important that the government be careful to avoid overregulating the institutional innovation that is presently occurring in forward pricing, contracting, insurance, risk-sharing, and marketing of products. For example, the current ban on trading in commodity options should be relaxed.

In general, it would be economically prudent and politically more feasible to ease into market approaches. Avoiding the abrupt cancellation of programs, we could gradually reduce real support prices. Indeed, a general trend in this direction is already under way. Today's programs, while costly, are not nearly as important to farmers in relation to their incomes as was the case during the 1950s and 1960s. In 1978, program gains to farm operators amounted to less than one-fifth of net farm income, and the 1979 and 1980 figures were undoubtedly less. Estimates for the 1950s and 1960s were more like one-third to one-half. Yet, farmers are substantially more prosperous today. And if we consider producers of particular commodities that have not been protected by programs, such as soybeans and hogs, there is no evidence that producers of them have suffered over the long term from their lack of program protection. Moreover, two important commodities, cotton and rice, moved in the 1970s from what had traditionally been a heavily regulated situation to a highly market-oriented program. No catastrophic or even mild problems for producers of these commodities have been traceable to the end of these highly interventionist programs.

In practice, prices and outputs are determined by a mixture of market and regulatory forces. This will continue to be true. The issue is one of emphasis: Do we want a market mechanism, with regulation only under clearly spelled-out and justified circumstances; or do we want a basically administrative pricing mechanism, with market forces permitted to rule only when they can be specifically justified and approved? Despite the increasing market orientation of policy for some commodities, there is evidence that we are moving toward the administrative view in the U.S.'s agricultural policy.

Recall the USDA's "expanded commitments" (quoted at the end of chapter 1) and the secretary of agriculture's interest in the "structure" of agriculture (chapter 5). More direct interventionist signals are given by the government's reflexive moves to stifle innovation in pricing, marketing, risk-sharing, and resource-procurement in a market context. To me, it is especially troubling that moves in this direction are being fostered by the seemingly more enlightened and publicly concerned segment of policy makers. Secretary Bergland's intentions to be broad-minded are reflected in the following description of his role in the market for milk: "Some consumers tend to want lower prices than will give producers the essential incentives for adequate supplies of milk in the long run. Some producers tend to want higher supports than are justified to provide an adequate supply. Secretaries of Agriculture must work to strike a balance."[9]

The striking aspect of this statement is the serene confidence that it is best for the nation that a market-pricing mechanism be replaced by governmental management of the price of milk. The conclusion of this book is that such confidence is mistaken and that the search for solutions to agriculture's problems through governmental intervention has been and will continue to be a costly delusion.

Appendix

Some points that are made in the text can be stated more precisely by means of supply/demand diagrams. A diagrammatic treatment is also helpful in showing how net social costs can be measured and how the effects of intervention in the commodity market depend on the market situation.

Figure A.1 compares the three types of intervention that are discussed in chapter 2. Panel *a* illustrates a market price support at Price $\overline{P}$ by means of governmental acquisition of excess supply at the support level. The quantity Q_D is commercial demand, and Q_S is the quantity offered for sale at the support price. The quantity $Q_S - Q_D$ represents what is acquired by the government.

Panel *b* shows the same price-support level, reached by restricting output to level Q_D. This eliminates any excess supply at the support price; it also avoids the accumulation of government-owned surpluses.

Panel *c* shows the same price-support level, guaranteed to producers by paying them the difference between $\overline{P}$ and the market price. The guarantee of $\overline{P}$ results in the production of Q_S. When placed on the market, the resulting price paid by consumers is P_c. Deficiency payments are $(\overline{P} - P_c) \times Q_S$, equal to the large hatched rectangle.

These diagrams illustrate the points that are made in chapter 4 about the gains to producers and consumers and the concept of the net social cost of intervention. Unrestricted production at the support price $\overline{P}$ creates additional economic rents—returns above those necessary to attract resources to the industry—equal to the upper hatched area in

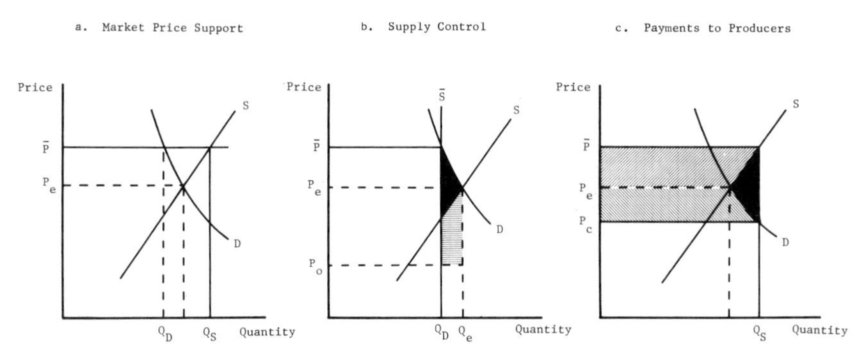

FIGURE A.1
THREE APPROACHES TO POLICY

panel *c*. The availability of quantity Q_S at price $\overline{P}$ (instead of the no-program price P_c) results in gains to consumers—increased consumers' surplus—equal to the lower hatched area. The net gains to consumers, producers, and taxpayers jointly is the sum of the two hatched areas, minus the rectangle representing government payments. Since the payments exceed the gains, there is a net social loss, measured by the shaded triangular area.

The large social costs for the set-aside and voluntary diversion programs in wheat and feed grains result from the idling of land. In panel *b* of figure A.1, the reduction of output from Q_e to Q_D eliminates products the value of which to consumers is the area under D between Q_D and Q_e. The resources that are released have value in alternative uses equal to the area under S between Q_D and Q_e. Thus, the net social cost is the shaded triangle in panel *b*. But for the set-aside programs, land typically does not go into its best alternative use. If the best permitted uses of diverted resources yield benefits P_o, then the net social costs of program are increased by the hatched area.

The point that producers are better off with demand-increasing rather than with supply-reducing programs can be shown with reference to panel *b*. Although the price, $\overline{P}$, is the same, the gain in economic rents is smaller when output is held to Q_D. Moreover, there is a loss of preprogram rents equal to the triangle bounded by S, $\overline{S}$, and the dotted line at P_e. Therefore, producers are better off in panel *c*. (The producers' gains in panel *a* are unclear, depending on what the government does in subsequent years with the commodities that it has acquired in order to support the price.)

The discussion in chapter 5 about disequilibrium is illustrated in figure A.2. The idea of overproduction of agricultural commodities can be expressed as disequilibrium at price P_o and quantity Q_o, with excess production of $Q_o - Q_e$. But it can also be said that P_o is a short-term equilibrium price at which the quantity actually produced along S_{sr} clears the market. If an error term were involved, this could also yield Q_o when Q_e was the intended output, but again, P_o would be the short-term equilibrium price. Thus, the term "disequilibrium" represents a constrained or short-run equilibrium outcome that is different from long-run equilibrium.

In this situation, measurement of net social costs is based on the (potential) long-run supply function, S_{lr}. Consider a support price for farmers at level $\overline{P}_f$, which has been achieved by deficiency payments of $(\overline{P}_f - \overline{P}_c) \times \overline{Q}$. The standard measure of social cost would be the small shaded triangle ECF. If, instead, we were to accept the irrelevance of S_{sr} and were to make our criterion for comparison the long-run equilibrium (price P_e and output Q_e), then the net social cost would be the much larger triangle ABC. This is a particular representation of the long-argued position that to the extent there is a farm problem, price sup-

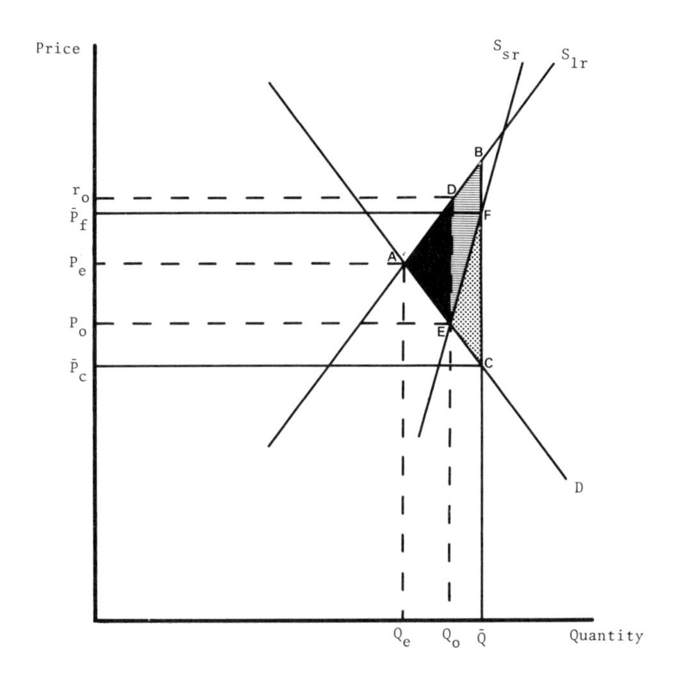

FIGURE A.2
SOCIAL COST UNDER DISEQUILIBRIUM

ports will only mask it and will tend to make worse the underlying misallocation of resources.

If the long-run supply curve is known, standard welfare economics can quantify the net social costs of failure to adjust to equilibrium. At the short-run equilibrium output Q_o, the long-run opportunity cost of resources used in farming is r_o, while the resources will yield farm products in agriculture valued at P_o. Therefore, by adjusting output to Q_e, there is a net social gain of the triangle ADE. A policy that would generate this adjustment in output—for example, by increasing labor mobility—would have gains valued at this amount (minus the costs of the policy).

Figure A.3 shows the demand curves for farm products as derived from the demand for the final product (food) through the actions of middlemen. The line labeled D_c results from competitive middlemen. Monopolistic middlemen buy less from farmers. Thus D_m lies to the left of D_c, resulting in farm production equal to Q_m instead of Q_c under competition. The resulting price received by farmers is P_m instead of the competitive P_c. The value of the farm products that are embodied in food products is R_m. Therefore, the returns to monopoly power are $(R_m - P_m)$ multiplied by Q_m. This is a transfer from farmers and

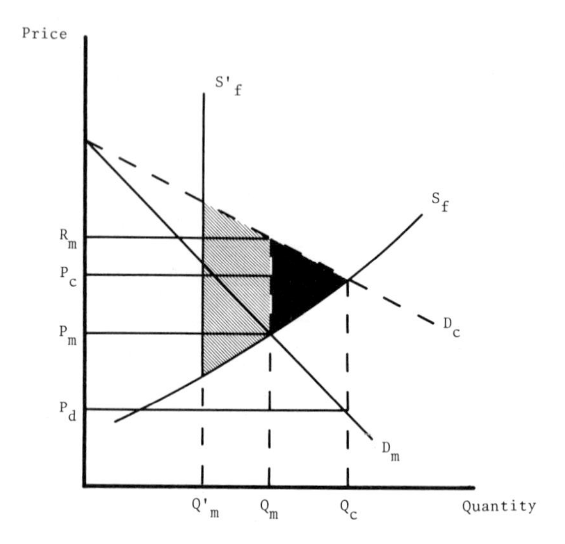

FIGURE A.3
THE EFFECTS OF MONOPOLY ON THE EVALUATION OF FARM PROGRAMS

consumers to middlemen; it involves a net social cost—a loss to farmers and consumers, which nobody gains—equal to the shaded triangular area.

Consider a policy that guarantees the competitive price P_c to farmers by means of deficiency payments with no restraints on production. The results would be production of Q_c, with the middlemen paying P_d and selling the product to consumers for P_c (on a farm basis, net of marketing costs). The triangle of net social cost would be eliminated. However, the middlemen's profits would be $(P_c - P_d)$ multiplied by Q_c, which would be larger than with no program. A program that made middlemen better off at the expense of taxpayers, who would essentially be subsidizing the middlemen's purchases of raw-product inputs, would be hard to accept politically.

In contrast, a production-control program would reduce output from Q_m to something like Q'_m. It might raise the price received by farmers (it will certainly do so for food monopolists, but the effect is indeterminate for monopsonists), and it will reduce the middlemen's profits. But the consumer price, determined by the intersection of D_c and S'_f, will be even higher than under unconstrained monopoly. And the net social cost will be *increased* by the hatched area.

The static supply/demand equilibrium analysis of farm programs is more generally applicable than is sometimes supposed, because it is adaptable to disequilibrium and imperfect competition along the lines introduced in this appendix. But this approach breaks down when production and demand are random variables. Mathematical tools of analysis then become indispensable. However, the net social costs of farm programs still exist and are measurable. Moreover, there is no reason to expect that the social costs of intervention will be smaller, in general, under uncertainty. For example, a support price that is below the mean price will have no social costs, indeed it will have no effects, in a static supply/demand analysis. But with random variation in prices, even a support price that is below the mean price will involve social costs. For an approach that carries through the idea of using price supports as put options in order to analyze programs under uncertainty, see my "Commodity Options for Agriculture," *American Journal of Agricultural Economics,* December, 1977, pp. 986–92.

Notes

CHAPTER 1

1. United States, Department of Commerce, Bureau of the Census, *Current Population Reports,* ser. P-27, no. 53: *Farm Population,* September, 1980.
2. United States, Department of the Treasury, Internal Revenue Service, *Statistics of Income—1977,* Individual Income Tax Returns (Washington, D.C., 1980).
3. D. W. Hughes, "Measuring Farm Operators' Nonfarm Wealth" (paper presented at a meeting of the American Agricultural Economics Association, Pullman, Washington, August, 1979).
4. United States, Department of Agriculture, Economics, Statistics, and Cooperatives Service, *Farm Real Estate Market Developments,* CD-83, July, 1978.
5. United States, House of Representatives committee report, quoted in *Washington Agricultural Record,* July 16, 1979.
6. United States, Department of Commerce, Bureau of the Census, *Current Population Reports,* ser. P-60, no. 120: *Money Income and Poverty Status of Families and Persons in the United States: 1978* (Washington, D.C., November, 1979).
7. USDA press release, text of speech by Secretary of Agriculture Bob Bergland, August 7, 1979.
8. H. L. Mencken, *Prejudices: A Selection* (New York: Vintage Books, 1958), p. 160.
9. USDA press release, remarks of Dale E. Hathaway, assistant secretary, USDA, January 9, 1979.
10. USDA press release, remarks of Secretary of Agriculture Bob Bergland, June 26, 1979.
11. *New York Times,* December 23, 1980.

CHAPTER 2

1. James Trager, *The Great Grain Robbery* (New York: Ballantine Books, 1975).
2. United States, House of Representatives, *Agriculture, Rural Development and Related Agencies Appropriations for 1980,* Hearings before the Subcommittee (of the Committee on Agriculture, Rural Development, and Related Agencies) on Appropriations, pt. 5, p. 355.
3. Ibid., p. 339.

4. United States, Senate, Committee on Agriculture, Nutrition, and Forestry, "Statement of Secretary of Agriculture Bob Bergland," January 24, 1979.
5. See, for example, D. McNicol, *Commodity Agreements and Price Stabilization* (Lexington, Mass.: Lexington Books, 1978).
6. *Wall Street Journal,* July 19, 1979.
7. It is possible that U.S. export restraint could be profitable to U.S. grain producers as well as to consumers. The principal criterion is that the demand for U.S. grain exports must be sufficiently inelastic relative to the U.S. domestic demand for grain so that reallocating a bushel of grain from world markets to the U.S. market will increase the average price of U.S. grains sold. Although it can be argued that the demand for U.S. grain exports was very inelastic from 1973 to 1975, it is unlikely to have been *this* inelastic, and grain producers were probably correct in believing that export restraints during this period would not have been in their economic interest.
8. United States, Department of Agriculture, "Outlook for U.S. Agricultural Exports," November 17, 1980.

CHAPTER 3

1. M. L. Cook, "The Sources, Limits, and Extent of Cooperative Market Power: Financial Law and Institutions," in *Agricultural Cooperatives and the Public Interest,* ed. B. W. Marion, North Central Regional Research Publication 256, University of Wisconsin-Madison (September, 1978).
2. United States, General Accounting Office, "Marketing Order Program," Report to Congress, April 23, 1976; United States, Department of Justice, "Milk Marketing," Report to the Task Group on Antitrust Immunities (January, 1977).
3. United States, Department of Agriculture, "Price Impacts of Federal Market Order Programs," Report of the Interagency Task Force (January, 1975).
4. See United States, Congressional Budget Office, "Consequences of Dairy Price Support Policy," Budget Issue Paper (March, 1979); J. W. Gruebele, "Effects of Removing the Dairy Price-Support Program," *Illinois Agricultural Economics,* July, 1978; D. Heien, "The Cost of the U.S. Dairy Price Support Program," *Review of Economics and Statistics,* February, 1977. For a summary of econometric results on elasticities of demands and supply, see H. L. Cook et al., *The Dairy Subsectors of American Agriculture,* North Central Regional Research Publication 257 (November, 1978).
5. Heien, "Cost of the U.S. Dairy Price Support Program," and Department of Justice, "Milk Marketing," especially the version edited by Paul W. MacAvoy as *Federal Milk Marketing Orders and Price*

Supports (Washington, D.C.: American Enterprise Institute for Public Policy Research, 1977).
6. See Margaret G. Reid, *Food for People* (New York: John Wiley, 1943).
7. USDA press release, April 12, 1979.
8. Related data are available from the reports required under the Agricultural Foreign Investment Disclosure Act of 1978. Data from the reports indicate that about 5.6 million acres were owned by foreigners (USDA Press Release, August 19, 1980).

CHAPTER 4

1. United States, Senate, "Food and Agriculture Act of 1977," Conference Report, September 9, 1977.
2. On estimating supply response under acreage-restraining programs, see J. P. Houck et al., "Analyzing the Impact of Government Programs on Crop Acreage," United States, Department of Agriculture, Economic Research Service, Technical Bulletin no. 1548, August, 1976. For details on the 1978 estimates, see Bruce Gardner, "U.S. Agricultural Policy—Its Economic Content, Context, and Rationale," in *U.S.-Japanese Agricultural Trade Relations,* ed. E. Castle and K. Hemmi (Washington, D.C.: Resources for the Future, in press).
3. For a more detailed analysis, see J. W. Freebairn and Gordon C. Rausser, "Effects of Changes in the Level of U.S. Beef Imports," *American Journal of Agricultural Economics,* November, 1975.
4. Joan S. Wallace, "Remarks to National Association of State Universities and Land Grant Colleges," USDA press release, July 25, 1979.
5. See Gordon Tullock, "The Welfare Costs of Tariffs, Monopolies, and Theft," *Western Economic Journal,* June, 1967; and Richard A. Posner, "The Social Costs of Monopoly and Regulation," *Journal of Political Economy,* August, 1975.

CHAPTER 5

1. See T. D. Wallace, "Measures of Social Costs of Agricultural Programs," *Journal of Farm Economics,* May, 1962; or for textbook discussion, see Edwin Mansfield, *Microeconomics: Theory and Application,* 1st ed. (New York: W. W. Norton, 1970).
2. George E. Brandow, "Policy for Commercial Agriculture, 1945–1971," in *A Survey of Agricultural Economics Literature,* vol. 1 (Minneapolis: University of Minnesota Press for the American Agricultural Economics Association, 1977), p. 271.
3. *Wall Street Journal,* August 17, 1979.
4. Donald R. Kaldor and William E. Saupe, "Estimates and Projections

of an Income-Efficient Commercial-Farm Industry in the North Central States," *Journal of Farm Economics,* August, 1966, pp. 578–96.

5. Frederick V. Waugh, "Does the Consumer Benefit from Price Instability?" *Quarterly Journal of Economics,* 1944, pp. 602–14.
6. R. Gray, V. Sorenson, and W. Cochrane, *Impact of Government Programs in the Potato Industry,* quoted in Willard W. Cochrane and Mary E. Ryan, *American Farm Policy, 1948–1973* (Minneapolis: University of Minnesota Press, 1976), p. 375.
7. Vernon W. Ruttan, "Program Analysis and Agricultural Policy," in *The Analysis and Evaluation of Public Expenditures,* United States, Congress, Joint Economic Committee (1969), vol. 3, p. 1140.
8. For a treatment in depth of this theme, see Don Paarlberg, *Farm and Food Policy* (Lincoln: University of Nebraska Press, 1980).
9. Bruce W. Marion et al., *The Food Retailing Industry* (New York: Praeger Publishers, 1979); R. C. Parker and J. M. Connor, "Estimates of Consumer Loss Due to Monopoly in the U.S. Food Manufacturing Industries," Working Paper WP-19, NC 117 (Madison: University of Wisconsin, September, 1978).
10. Remarks of Secretary of Agriculture Bob Bergland before the National Farmers Union Convention, March 12, 1979, as released by the USDA.
11. Ibid.
12. Stanley Aronowitz, *Food, Shelter, and the American Dream* (New York: Seabury Press, 1974), p. 103.
13. A 1978 USDA survey estimated that 21.2 million acres of farm and ranch land were owned by 37,000 nonfamily corporations. This is larger than the census estimate but still represents only 2.4 percent of total U.S. farm acreage. See USDA, Economics, Statistics, and Cooperative Services, "Who Owns the Land?" no. 70 (September, 1979).
14. Michael Perelman, *Farming for Profit in a Hungry World* (New York: Universe Books, 1977). See also C. Lerza and M. Jacobson, eds., *Food for People, Not for Profit* (New York: Ballantine Books, 1975).
15. Bruce Gardner, "Farm Population Decline and the Income of Rural Families," *American Journal of Agricultural Economics,* August, 1974.
16. J. Madden and D. Brewster, eds., *A Philosopher among Economists: Selected Works of John M. Brewster* (Philadelphia: J. T. Murphy Co., 1979), p. 151.

CHAPTER 6

1. For further discussion of some of these points, see John Schnittker, "The 1972–73 Food Price Spiral," *Brookings Papers on Economic Activity* 2 (1973): 498–514.

2. Cochrane and Ryan, *American Farm Policy,* pp. 391–92.
3. D. Gale Johnson, "International Food Security," in *International Food Policy Issues* (Washington, D.C.: U.S. Government Printing Office, 1977).
4. For further discussion, see Bruce L. Gardner, *Optimal Stockpiling of Grain* (Lexington, Mass.: Lexington Books, 1979), chap. 6; or J. Sharples, "An Alternative Farmer Reserve Program," USDA-ESCS, Purdue University, April, 1980.
5. See Milton Friedman, *A Theory of the Consumption Function* (Princeton, N.J.: Princeton University Press, 1957), pp. 61–64; and Margaret G. Reid, "Effect of Income Concept upon Expenditure Curves of Farm Families," in *Studies in Income and Wealth,* Conference on Research on Income and Wealth, vol. 15 (New York: National Bureau of Economic Research, 1952), pp. 133–74.
6. United States, Senate, Hearings, January 23, 1978, as quoted in Gerald McCathern, *From the White House to the Hoosegow* (Canyon, Texas: Staked Plains Press, 1978), pp. 134–36.
7. I. Lukinov, "The Methodology of Forming Prices of Farm Produce, History of Price Formation in the USSR," Papers and Reports, International Conference of Agricultural Economists, 1970, p. 240.
8. See James P. Houck, "Agricultural Trade: Protectionism, Policy, and the Tokyo/Geneva Negotiating Round," *American Journal of Agricultural Economics,* December, 1979, pp. 860–73.
9. USDA press release, Remarks by Howard Hjort, director of economics, policy analysis and budget, April 2, 1980.

Bibliography of Works Cited

Aronowitz, Stanley. *Food, Shelter, and the American Dream.* New York: Seabury Press, 1974.

Brandow, George E. "Policy for Commercial Agriculture, 1945–1971." In *A Survey of Agricultural Economics Literature,* vol. 1. Minneapolis: University of Minnesota Press for the American Agricultural Economics Association, 1977.

Cochrane, Willard W., and Ryan, Mary E. *American Farm Policy, 1948–1973.* University of Minnesota Press, 1976.

Cook, H. L., et. al. *The Dairy Subsectors of American Agriculture.* North Central Regional Research Publication 257, November, 1978.

Cook, M. L. "The Sources, Limits, and Extent of Cooperative Market Power: Financial Law and Institutions." In *Agricultural Cooperatives and the Public Interest,* ed. B. W. Marion. North Central Regional Research Publication 256. Madison: University of Wisconsin-Madison, September, 1978.

Freebairn, J. W., and Rausser, Gordon C. "Effects of Changes in the Level of U.S. Beef Imports." *American Journal of Agricultural Economics,* November, 1975.

Friedman, Milton. *A Theory of the Consumption Function.* Princeton, N.J.: Princeton University Press, 1957.

Gardner, Bruce L. "Farm Population Decline and the Income of Rural Families." *American Journal of Agricultural Economics,* August, 1974.

———. *Optimal Stockpiling of Grain.* Lexington, Mass.: Lexington Books, 1979.

———. "U.S. Agricultural Policy—Its Economic Content, Context, and Rationale." In *U.S.-Japanese Agricultural Trade Relations,* ed. E. Castle and K. Hemmi. Washington, D.C.: Resources for the Future, in press.

Gruebele, J. W. "Effects of Removing the Dairy Price-Support Program." *Illinois Agricultural Economics,* July, 1978.

Heien, D. "The Cost of the U.S. Dairy Price Support Program." *Review of Economics and Statistics,* February, 1977.

Houck, James P. "Agricultural Trade: Protectionism, Policy, and the Tokyo/Geneva Negotiating Round." *American Journal of Agricultural Economics,* December, 1979, pp. 860–73.

Houck, J. P., et al. "Analyzing the Impact of Government Programs on Crop Acreage." United States, Department of Agriculture, Economic Research Service, Technical Bulletin no. 1548, August, 1976.

Hughes, D. W. "Measuring Farm Operators' Nonfarm Wealth," paper presented at a meeting of the American Agricultural Economics Association, Pullman, Washington, August, 1979.

Kaldor, Donald R., and Saupe, William E. "Estimates and Projections of an Income-Efficient Commercial-Farm Industry in the North Central States." *Journal of Farm Economics,* August, 1966, pp. 578–96.

Lerza, C., and Jacobson, M., eds. *Food for People, Not for Profit.* New York: Balantine Books, 1975.

MacAvoy, Paul W., ed. *Federal Milk Marketing Orders and Price Supports.* Washington, D.C.: American Enterprise Institute for Public Policy Research, 1977.

McCathern, Gerald. *From the White House to the Hoosegow.* Canyon, Texas: Staked Plains Press, 1978.

McNicol, D. *Commodity Agreements and Price Stabilization.* Lexington, Mass.: Lexington Books, 1978.

Madden, J., and Brewster, D., eds. *A Philosopher among Economists: Selected Works of John M. Brewster.* Philadelphia: J. T. Murphy Co., 1979.

Marion, Bruce W., et al. *The Food Retailing Industry.* New York: Praeger Publishers, 1979.

Paarlberg, Don. *Farm and Food Policy.* Lincoln: University of Nebraska Press, 1980.

Parker, R. C., and Connor, J. M. "Estimates of Consumer Loss Due to Monopoly in the U.S. Food Manufacturing Industries." Working Paper WP-19, NC 117. Madison: University of Wisconsin, September, 1978.

Perelman, Michael. *Farming for Profit in a Hungry World.* New York: Universe Books, 1977.

Posner, Richard A. "The Social Costs of Monopoly and Regulation." *Journal of Political Economy,* August, 1975.

Reid, Margaret G. "Effect of Income Concept upon Expenditure Curves of Farm Families." In *Studies in Income and Wealth,* vol. 15, pp. 133–74. New York: National Bureau of Economic Research, 1952. Conference on Research on Income and Wealth, November, 1952.

———. *Food for People.* New York: John Wiley, 1943.

Ruttan, Vernon W. "Program Analysis and Agricultural Policy." In United States, Congress, Joint Economic Committee, *The Analysis and Evaluation of Public Expenditures.* Vol. 3, 1969.

Schnittker, John. "The 1972–73 Food Price Spiral." *Brookings Papers on Economic Activity* 2 (1973): 498–514.

Sharples, J. "An Alternative Farmer Reserve Program." USDA-ESCS, Purdue University, April, 1980.

Trager, James. *The Great Grain Robbery.* New York: Ballantine Books, 1975.

Tullock, Gordon. "The Welfare Costs of Tariffs, Monopolies, and Theft." *Western Economic Journal,* June, 1967.

United States, Congressional Budget Office. "Consequences of Dairy Price Support Policy." Budget Issue Paper, March, 1979.

United States, Department of Agriculture. "Price Impacts of Federal Market Order Programs." Report of the Interagency Task Force, January, 1975.

United States, Department of Agriculture, Economics, Statistics, and Cooperatives Service. *Farm Real Estate Market Developments,* CD-83, July, 1978.

United States, Department of Agriculture, Economics, Statistics, and Cooperative Services. "Who Owns the Land?" No. 70, September, 1979.

United States, Department of Commerce, Bureau of the Census. *Current Population Reports,* ser. P-27, no. 53: *Farm Population,* September, 1980.

United States, Department of Commerce, Bureau of the Census. *Current Population Reports,* ser. P-60, no. 120: *Money Income and Poverty Status of Families and Persons in the United States: 1978.* Washington, D.C., November, 1979.

United States, Department of Justice. "Milk Marketing." Report to the Task Group on Antitrust Immunities, January, 1977.

United States, Department of the Treasury, Internal Revenue Service, *Statistics of Income—1977,* Individal Income Tax Returns. Washington, D.C., 1980.

United States, General Accounting Office. "Marketing Order Program," Report to Congress, April 23, 1976.

Wallace, T. D. "Measures of Social Costs of Agricultural Programs." *Journal of Farm Economics,* May, 1962.

Waugh, Frederick V. "Does the Consumer Benefit from Price Instability?" *Quarterly Journal of Economics,* 1944, pp. 602–14.

Index